STOMP YER CROC!

How to bring real & lasting
IMPROVEMENT
to your organization...and have
FUN doing it!

Ralph Williams

Foreword by **Mike Konrad, Ph.D.**
Cooliemon® Press
Pittsburgh, PA

Cooliemon® Press
Pittsburgh, PA
Copyright © 2008 by Ralph Williams
All rights reserved. This book or parts thereof may not be reproduced in any form without permission.

Noel Knowall™ and Minkee the Monkey™ are Cooliemon trademarks.

Published simultaneously in Canada.
Williams, Ralph
Stomp Yer Croc!™: *How to bring real & lasting* IMPROVEMENT *to your organization... and have* FUN *doing it!*
First edition
p. cm.
ISBN 978-0-9720024-7-9
For more information about *Stomp Yer Croc!*
visit www.cooliemon.com

Editing, interior & jacket design, and typesetting
by
Schmidt Kaye & Co.
Houston, Texas

Printed in Canada

This book is dedicated to my Mum and Dad.

I have the most incredible parents a person could possibly have. I will always cherish my childhood memories of your love and support over the years. The values you instilled in me have flowed into my life's work. Without your patient hands and guidance, I never would have had the greater vision of life's possible opportunities that you passed on to me.

Contents

INTRODUCTION — 1

FOREWORD — 7

THE DEBRIEF — 11

Chapter 1 ~ THE HUNTER'S TALE — 15

Chapter 2 ~ THE NATURE OF THE HUNT — 35

Chapter 3 ~ A FRESH HUNT — 55

Chapter 4 ~ GAUGING PROGRESS — 75

Chapter 5 ~ STOMPED CROCS! — 91

THE POST-GAME MEETING — 107

SOME FINAL THOUGHTS — 131

Not to be too cheeky about it, but if you bought this book expecting an instruction manual on crocodile hunting, I fear you're in for a surprise. In truth, no actual crocodiles were harmed in the writing of this book, nor do I advocate any such activity. Stomping crocs, in the context of this book, is purely a figure of speech.

On the other hand, if your interest is in the perils and pleasures of overcoming the problems that menace your company, this book will put you in very familiar territory. The characters and events of the story represent the basic aspects of these problems. I feel certain that you will see members of your own organization in some of the characters. You will also no doubt see the similarity between situations the characters encountered on the hunt and those you face in the course of doing business. Some are people-related problems, and some are technology-based, while others focus on processes and less-than-effective ways of doing business.

More Than Problems

Stomp Yer Croc™, however, is about more than just addressing problems. It is about a revitalized approach to creating solutions. The tale suggests specific successful approaches proposed by Shewhart, Feigenbaum, and Juran that have literally disappeared from use over the past half a century in all areas of business except manufacturing.

Walter Shewhart combined creative management thinking with statistical analysis that focused on measure-based accountability. His work was summarized in 1931 in his book, *Economic Control of Quality of Manufactured Products*.

In 1951, Armand Feigenbaum stated that up to 40% of the capacity of manufacturing plants was being wasted. At the time, it was thought by some to be an unbelievable figure; even today, some managers appear not to realize that such high levels of inefficiency are common. Joseph Juran later echoed Feigenbaum by stating that the cost of poor quality in any company was between 20% and 40% of revenue.

The Star Organizational Improvement (Star OI™) method outlined in *Stomp Yer Croc*™ uses whatever model, standard, or business improvement approach that best fits your organization. It is not dependent on, nor does it promote, any specific business improvement method or standard, such as:

- Capability Maturity Model Integration (CMMI®)
- Six Sigma
- International Standards Organization (ISO)
- Baldridge National Quality Program
- Institute of Electrical and Electronics Engineers (IEEE)
- Information Technology Infrastructure Library (ITIL®)
- Business Process Re-engineering (BPR)
- Total Quality Management (TQM)

All of these methods and standards, to a greater or lesser extent, reflect the seminal work of Shewhart, Feigenbaum, Deming, and Juran. The question you, the individual business owner or manager, must ask is whether your improvement program is yielding measurable, ongoing, and institutionalized financial rewards. For example…

- Are you achieving the kinds of success that you hoped for or envisioned?
- Have you started an improvement journey and not finished it?
- Have you only done enough to earn and maintain a certification that always costs but does not save money?
- Do you feel that your firm will never truly reap the full benefits of improvement?

If any of these situations sounds familiar, you are far from being alone. There are many great techniques out there to help you, and in this book you'll discover how to make the most of those techniques. As a Software Engineering Institute (SEI) Transition Partner and authorized Lead Appraiser, I know the strengths of the CMMI®Model, as well as the positive contributions of ISO, Six Sigma, Baldrige, and ITIL. Star OI™ complements these models and standards, leverages their strengths, and enhances their primary goal of improving organizational performance. The last chapter in this book, "Some Final Thoughts," cites the kinds of consistent and ongoing return on investment that your organization should enjoy year after year.

But I'm getting ahead of myself. Let's begin at the beginning…

About the Story

The beginning of the *Stomp Yer Croc*™ story is an analogy about how an otherwise successful organization experiences problems during an important project. During his quest to reach the Valley of Gold, Jason Hunter, a well-known and successful hunter, runs into:

- Communication problems
- Training issues
- Inadequate resources to perform the work
- Unclear lines of responsibility
- Failure to make realistic commitments
- Turf wars
- Insufficient planning
- Lack of stakeholder involvement
- Resistance to change

Determined to fix these problems on the next hunt, Jason solicits help from a personal business improvement coach named Alex, who has a reputation as one of the world's best hunters. Alex's greatest strengths are not just his extensive knowledge and skills, but the way in which he works with Jason and his team.

As a result of what happens during the second hunt, everyone sees the value in how Alex had them plan, track, and measure their own activities. They also see how his "people skills" are applied to individual and situational areas of potential improvement.

Of greater importance, the "value" of the respective skills of the individuals on the hunt becomes clear. With help from Alex, Jason's team gains significant results on the second quest. As the fable ends, an edified Jason is extremely confident because he knows how to create desired results!

Acknowledgments

First, I would like to thank my dear friend and wonderful colleague, Jack Maples, who helped put life into my business improvement method and characters by working with me on the *Stomp Yer Croc*™ story. Thank you, Jack, for all of your hard work, as well as your polishing, reviews, comments, edits, and other contributions. Your experience as the author of *Reconstructed Yankee* paid off during this venture.

I also must add a very special thank you to Ron Kaye and Connie Schmidt, who have added so much to the story. Their efforts increased the clarity of the tale and developed the characters in ways that greatly improved the flow of the plot and the points that the story conveys. In addition, Ron and Connie designed the wonderful dust jacket for *Stomp Yer Croc*™ that illustrates the themes and symbols represented in the book, and designed the interior in a manner that significantly enhanced the book's visual appeal.

My thanks also must go to my illustrator, Mike Tanoory, who worked diligently to create the cartoons. Mike, it was a joy to see the way you made my characters come alive.

For my family and friends, I cannot do what I do without your loving patience and support. My thanks to you is everlasting.

And For You, the Reader…

I hope you enjoy reading this story as much as we enjoyed putting it together. It is meant to be a refreshing way to look at typical business issues that we see in all kinds of organizations, regardless of size or industry. Thus, as I noted earlier, some of the scenes should resonate quite closely with your experiences in the workplace.

By the way, I have referred to groups of crocodiles as croc herds. Technically, they are called "floats." My decision to refer to them as herds was conscious. It is my hope that those who prefer pristine uses of words will forgive me this exercise in creative license.

More importantly, I hope that, *Stomp Yer Croc*™ will stimulate change in your organization by guiding the way to a new perspective on how even highly successful companies can improve. Included herein are a number of hard-learned key lessons that you can apply in your organization today.

Ralph Williams

> *"Problems cannot be solved at the same level of thinking that created them."*
> ~ Albert Einstein

Working with various organizations continues to be a wonderful profession. Meeting so many great folks, in both private-sector businesses and the government world, and working alongside them has been a continual inspiration. My goal is to help these organizations leap over their hurdles, address their problems, and resolve their challenges in a proactive fashion – to stomp their own particular "crocs" and find the gold they deserve.

Stomp Yer Croc!

foreword

I was genuinely pleased when Ralph Williams asked me to write the Foreword to this book, because *Stomp Yer Croc*™ is a unique story with valuable lessons for people in business everywhere. It tells the tale of Jason Hunter, an adventurer extraordinaire who, like all of us, is human and capable of errors of judgment and overconfidence. Also, like all of us, he is capable of learning from his mistakes.

The story starts with Jason planning and then making an initial safari into the legendary Valley of Gold. Jason has been through his share of "projects" before, and he realizes that hunting trips, much like "projects" in the business world, require planning, staffing, and resources. He also knows that, just as in "real" life, unexpected challenges are often encountered on the hunt, and the entire enterprise will have aspects of both failure and success.

Stomp Yer Croc™ also introduces other wonderful characters and elements that help bring out the points of the story, including:

- **Alex**, a world-renowned hunter in his own right, who can speak bluntly and point Jason to a better path
- **Noel**, a team lead and experienced senior hunter, who knows it all and resists changes, but who also grows to become a valuable member of the team
- **Minkee the Monkey and his friends**, who are the pet projects and distractions that derail the team's efforts
- **Chunky chickens, ox carts, and SUVs**, which are the outdated and bloated processes and ineffective "short cuts" that cost time and money
- **Roadblocks and walls** that force re-scheduling and a re-direction of efforts, to the detriment of product and service quality
- Then, of course, there are the **individual crocs, the croc herds, and their nests**, which represent the

problems, contributing problems, and root causes of problems, respectively, that plague us all. Led by **King Croc – the Father of All Problems** – crocs are the myriad undiscovered sources of waste and inefficiency that limit efficiency, productivity, and profitability.

More often than not, we fail to see these crocs because we have come to accept them as just "part of the business environment." *Stomp Yer Croc*™ also makes the point that these are not just problems with process or quality; rather, our crocs come in many forms, and require a variety of types of solutions.

What makes the story compelling and useful is that like any good business owner, Jason is determined to do better on each successive expedition. Thus, he begins an internal soul-searching journey that is overlaid on a second safari to the Valley of Gold – a journey of learning how to manage better – and that's where the story hits the "real gold." He uncovers the nuggets of insight that we "mine" through personal reflection and by learning from others. By mining these nuggets, Jason learns to avoid repeating mistakes and improves the success of his endeavors.

In summary, this book is about the most important journey of all, which is simply the understanding of self, of our environment, and of what we can do to help us better succeed. In a sense, each of us is "Jason Hunter" as we reflect on how to do better on our next safari into our own unique Valley of Gold.

Reading *Stomp Yer Croc*™ caused me to reflect on the start of my own personal "safari." My mother was French and my father was a Polish-American who grew up in the Philippines, and whose education was disrupted because of World War II. Both of my parents placed a strong emphasis on their children's education.

Even in my early childhood, I discovered that I possessed a passion for mathematics. As I pursued my education, I was also blessed with some great teachers and mentors. However, after receiving my Ph.D., I discovered to my dismay that I hated actually being an academic mathematician. So much time was devoted to preparing articles for publication that it literally took the joy out of the mathematics that I dearly loved.

I still desired to use my "talents" in some way, in part to "pay it forward" for the sacrifices made by others on my behalf. This led me on a new quest. My initial explorations of computer science took me into the field of artificial intelligence, and then to systems and software engineering.

Along the way, I added to my brief experience of academia the diverse worlds of private industry, government contracting, and a commercial startup with aspirations of creating the "next Lotus 1-2-3." Ultimately, my trek led me to my own Valley of Gold as the manager of one of the most successful teams within the Software Engineering Institute (SEI), the Capability Maturity Modeling® Team, and a product known as the Capability Maturity Model® Integration(CMMI®). Ralph's approach to organizational improvement, embedded in *Stomp Yer Croc*™, focuses on how to make theoretical best practices, such as those found within CMMI® and ITIL, "real" in project management, product development, and overall organizational improvement.

Just as the mission of the SEI is to advance software engineering and related disciplines in order to ensure the development and operation of systems with predictable and improved cost, schedule, and quality, Ralph's methods and techniques aim to advance the effectiveness and scope of an organization's improvement efforts to accomplish broader outcomes.

In this context, *Stomp Yer Croc*™ offers value because it:

- Is an allegory about true professionalism: striving to do your best through understanding your capabilities and how they can be improved
- Highlights many key principles found and elaborated on within the CMMI®, such as project planning, risk management, measurement and analysis, and causal analysis and resolution
- Illustrates the kinds of situations in which real-life managers find themselves, and the lessons that can be drawn from such situations
- Focuses not merely upon gaining organization-wide improvements that reduce cost and improve profit, but on maintaining and even enhancing those improvements over the long term

Stomp Yer Croc!

- Is a fun and stimulating read about an area so important to us: executing each project to the best of our organization's abilities, and achieving even better results than on our previous efforts.

Having known Ralph for about ten years, I consider him a trusted friend and astute colleague, with unique insights into what makes businesses succeed. Ralph's expertise and focus are not limited to learning what makes a business "tick"; he is also driven to share his experience with others, and to get them on a path to greater improvements.

Ralph wants organizations to understand what it takes to succeed in their business endeavors. Toward that objective, I feel certain that *Stomp Yer Croc*™ will prove invaluable in helping business people and government workers obtain this greater understanding. To further demonstrate his ideas, Ralph has embellished his unique storytelling style with more than forty color cartoons and illustrations . The result is a blend of a business fable and a graphic novel that is a visual delight, as well as a powerful learning tool.

May your own safaris to the Valley of Gold be as stimulating and fun as was Jason Hunter's.

Mike Konrad
Software Engineering Institute
Carnegie Mellon University
January 5, 2008

the debrief

Carson arrived in the conference room early, carrying with him copies of a few materials for the other company executives. He set up his PC and plugged it into the projector.

During the previous week, Carson had attended a seminar series for Chief Administrative Officers. Though the series had been a relaxing experience overall, one seminar had captured his attention. It was the reason for this meeting, and he was frustrated because the others hadn't arrived yet. "Patience," he thought. "They're busy, but they'll be here."

Helen, the Chief Financial Officer, who was known as Coco, came in first. "Hey, Car! How was your vacation? Heard you played a little golf."

"Sure did. And I shot two over par," he replied. "With my handicap, it was a good round."

"Right. You're having fun and we're back here slaving away."

"It wasn't all fun. As a matter of fact, when Buddy and Marsha get here, you'll see that I actually came back with something useful."

Carson was somewhat surprised when Marsha, the CEO, came in as he spoke. She was usually the last to arrive at these meetings. "Good afternoon, Coco. Welcome back, Car," she said. "First time I've seen you today."

Buddy was the last to arrive. "Sorry, Car, we had a hiccup when backing up the network. Hope I'm not too late."

"Just getting ready to start, Buddy," Carson answered.

"It's already after three. Hope this is quick," Coco interjected. "My spareribs are par cooked, but I need to get the smoker started to finish them off for the tailgate party."

"I know," Carson said. "We all have season tickets, and I need to swing by the house for my stuff. A Monday night game at home, and we're going to kick some serious tail."

Marsha said, "Then, let's get going. Why the fancy debrief instead of an e-mail summary, Car?"

11

Stomp Yer Croc!

Carson stood up to speak. "Well, I'll be sending you all my customary report on most of the sessions, but there was one seminar that was special. At least it was a little different from the run-of-the-mill bits and pieces we usually get from these things."

"So are you going to share, or do show and tell?" Coco needled him, albeit with a smile.

"I'm just going to share a little now. The seminar ran for three hours, and we don't have enough time for a detailed presentation today. There are, however, a few high points I need to tell you about, and then we'll go to the game. I'd like to meet again tomorrow, after you've read this little story."

Carson passed around the copies to Marsha, Buddy, and Coco. "Don't start reading now. You'll get caught up in it, and we'll spend the game talking about it instead of talking football. After all, we do need to keep our priorities in order."

Everyone laughed at this remark, and then disregarding Carson's advice, glanced down at their copies. "A hunting story!" Marsha exclaimed. "I thought this was going to be something useful."

"Patience, Marsha. That's what a lot of people in the seminar said. Let me explain."

"Please do," Coco said. "This seems silly."

Carson had expected this kind of response. Undaunted by their cynicism, he explained to them that the first hour of the seminar had been really rough. Several people had even threatened to walk out, but were encouraged to stay.

"Give Car a chance," Buddy said. "If he thinks it's worth it, we should at least listen. We're not just in business together. We're friends, too."

"You shouldn't need more than an hour to read the story," Carson added. It's a pretty simple tale with some interesting parallels."

"Parallels to what?" Coco asked.

"Us," Carson replied. "And to our business. It's also about improving or getting better."

"We're already pretty good," Marsha said. "Look at our stocks and our profits. We don't have much room for improvement. I'm more interested in increasing our market share."

"So was Jason, the main character in the story, " Carson said. "Read the story, and you'll get the gist of it. I just want to share some cautions before we leave for the game."

"That seems odd," Buddy interrupted. "Why would the story need any cautions?"

"Because the main character, Jason, changes as the story progresses. My first impression was that the story was slamming executives and their managers, but that wasn't the point. After the first chapter, folks at the conference wasted a lot of time justifying Jason's actions, rather than actually thinking clearly about them."

Coco laughed out loud. "They got defensive. What a hoot! I love seeing folks on the hot seat. So, the story gets a little too close to the truth. What else?"

Coco had made Carson smile, too. "Don't overanalyze the symbols. Just read the story for the sheer pleasure of it. You'll see a lot of characters and actions we've seen here. I even saw myself a little too closely for comfort."

"So, Car, what happens tomorrow?" Buddy asked.

Stomp Yer Croc!

"I'll lead the discussion. I'm interested in seeing what each of you gets out of it. But right now, I'm ready for those ribs, Coco."

"And for a high-scoring game," Marsha added. "Are you ready for some football? Let's get going."

The Hunter's Tale

Jason Hunter was well known and respected all over the world for his successful safaris. In fact, he had turned what had once been a hobby, and then an overriding passion, into a lucrative profession; his company, Hunter Adventure Safaris, was one of the most successful in the industry. Now, however, he faced a challenge that would test even his formidable talents. The report on his desk described a Valley of Gold, filled with riches beyond measure. It was to this valley that Jason was planning an expedition.

The valley, however, was not a typical one; it was located in an extremely inaccessible part of the world, its entrance surrounded by a water-logged swamp, choked with undergrowth and heavily forested. Impassable mountains with sheer cliffs stood formidable guard over its other sides.

Others had attempted the long and difficult trip before, but no previous safari had succeeded completely in penetrating the borders of the valley. Many had failed completely. Entering the valley by helicopter or parachute drop was out of the question, due to the extreme turbulence that formed beneath the towering cliffs. Those who had tried this fast route were the ones who had failed most miserably. Jason realized that the rarely used and overgrown "road" through the swamp remained his only option.

To say that the dirt road into the valley was poorly maintained was a gross understatement. In most places, it was little more than an overgrown path that had long ago been reclaimed by the jungle. The swamp through which it twisted was hot and soggy, a perfect habitat for all types of menacing plants, reptiles, and animals. To no one in particular, Jason muttered, "Great! Crocodiles, snakes, and bugs!"

Stomp Yer Croc!

The report included local myths about the valley and the swamp. According to the report, the local villagers had made it taboo to enter the swamp because of the grave dangers that had faced those who had tried. The village lore included many stories about one large and very ancient croc that would devour anyone who came his way. They called him "King Croc," and according to legend, he was some kind of supernatural creature whose sole obsessive purpose was to guard the valley's riches. In spite of these obstacles – both real and mythical – Jason Hunter was confident that with the combination of the right people and the right methods, his safari, unlike others, would succeed.

Jason felt that the best course was to put together a select team of experienced hunters, who would, in turn, lead smaller safari teams in pursuit of the gold. Jason felt that with him in command, these teams could triumph over any obstacles. After all, he reasoned, the right people would have the experience and tools required for meeting any challenges they would face. That was true in any business, he thought, and his was no exception.

After gathering all of the information he could find, Jason called a meeting of his Safari Leads. Hours of meetings turned into days, then weeks. Despite many heated discussions, they all finally agreed on what they believed to be the best way to conduct the expedition. Jason and his Safari Leads thought they had created the ultimate Safari Plan.

Next, Jason turned his attentions to the task of financing the expedition. Fortunately, he was able to raise the money he needed, because bankers and financiers alike knew his reputation and believed in his ability. Once the financing was in place, Jason gave his Safari Leads the authority and money that they needed to hire their teams and buy their own equipment. While the Safari Plan indicated what supplies they would need for the journey, the Safari Leads had considerable leeway in deciding exactly what to purchase.

Jason had every confidence that no detail was being overlooked. He wasn't even worried about the crocs. There were plenty of road maps to guide the way, as well as satellite telemetry to help determine exactly where they were at all times, and, equally as important, where those crocs might be. I *will definitely be prepared*! Jason said to himself.

I will definitely be prepared!

Following weeks of meetings and planning, the teams were finally assembled and equipped. Jason had set the date and time to gather for the next pre-departure meeting, two weeks before the journey was to begin, and made certain that all personnel were informed and would be in attendance. That way, any problems could be resolved before the teams entered the swamp. Everyone agreed that Supply City, the small city near the road into the swamp, was the best place to meet. Although there was a small town closer to the swamp, it lacked the extra supplies they might need.

This is a well planned safari! Jason thought. *Best one I'll ever lead. We have the best people, too. And tomorrow I'll be on my way to Supply City. That gold is ours for the taking. We'll march through the swamp, stomp all the crocs, get to the valley, and be back in four months.* Another thought quickly followed that one: *Better be, or the bank will eat our lunch!*

Stomp Yer Croc!

Jason read the Safari Plan one last time, reviewing every detail with care. He went over his personal checklist again too. "Yep," he said aloud. "It's all there. We're most certainly ready." Setting his checklist aside, he leaned back in his chair, a satisfied smile on his face as he offered a self-congratulatory toast to the plans that lay on the large desk before him.

Three days later, after a long flight, a train ride, and a punishing journey by bus, Jason finally arrived in Supply City. He was anxious to meet his Safari Leads again, if for no other reason than to see how they had used the funds, and to get a feel for the people they had hired. More than anything, he was anxious to get the show on the road. After all, this expedition had been in the planning stage for three full months. Now, it was finally time to act, and to actually do what Jason did best.

Gathering For The Hunt

As he walked through the Base Camp, Jason knew that something wasn't quite right. He didn't just sense it. He knew it. Overall, things looked right on the surface. There were tents, supplies, and vehicles in abundance. Other things, however, looked out of place to him. The Jeeps looked sturdy and new, but why was there an armored personnel carrier? Then he saw the luxury SUVs. "My word! We agreed on transportation... but SUVs?"

What really took him by surprise, though, were the ox carts. Some of the carts were loaded with bags of grain, and one had several crates of chickens...big, plump chickens...chunky chickens. *Now, why on earth do we need chickens?* Jason thought to himself. *For that matter, why are we using ox carts? And what's with all this grain? Is it for the oxen?* His team leaders had better have some answers...

Ox carts?! Chunky chickens?!

Jason also noticed that each Safari Lead had hired several people to build his team. But there appeared to be a wide variation in the way each team was structured. Some were made up entirely of worker bees, while others had multiple hierarchies with sub-team leaders. *How are they going to get the work done*? Jason asked himself, growing increasingly concerned. *What surprises am I not seeing yet*? His ruminations were interrupted as he approached the Headquarters Tent, where he heard an increasingly heated argument in progress. As he entered the tent, he found the Safari Leads, who instantly fell silent. "What's the problem?" Jason asked.

"Nothing, Mr. Hunter," one of the Leads answered. "A minor logistical issue, actually. Some of us think the ox carts should be in the front because they won't break down and prevent the rest of the group from proceeding. Others are of the opinion that they'll just slow us down, because the Jeeps and SUVs move faster."

Jason was amazed. "Gentlemen, most of the team will be travelling on foot. No matter how fast the Jeeps and SUVs move, we can't go any faster than people – and, for that matter, oxen – can walk. Which brings me to my real question: why have ox carts or SUVs at all? The Jeeps are the most practical."

Another one of the Safari Leads spoke up, "Sir, I've been hunting for more than twenty years. I've always taken ox carts. The new and improved ones don't get stuck in the mud like the Jeeps can. Besides, they don't need fuel like the Jeeps and SUVs. Oxen just graze on whatever grasses are available along the road."

"So then, what is all that grain for, if not to feed the oxen?" Jason asked.

With a broad smile on his face, another Safari Lead answered. "It's for biscuits to feed our team members."

"Well, I don't know much about making biscuits," Jason admitted, "but it seems that you'd have to convert the grain into dough first. How do you propose to do that?"

"Got it covered," replied the Lead. "We're bringing a couple of portable mills to make flour – everything we need to make the most nutritious and delicious biscuits you've ever had. Oh, and we're bringing an experienced trail cook with us too. I can assure you that our team will be very well fed."

Before Jason could comment further, the youngest Safari Lead added, "And as for my SUVs – they have winches and the latest GPS technology. We won't get lost, and they can pull the Jeeps out of the mud if they get stuck."

"All right – granted, the ox carts and SUVs and flour mills weren't specified in the plan," Jason conceded. He knew it was best to keep the peace and avoid offending anyone before the trip even began. He did his best to calm his crew down, acknowledging the reasoning behind each of their transportation choices. They obviously had some emotional stake in these matters. "I can see why each of you made your choices of appropriate transportation," he continued. "But tell me, what's with all of these chickens?"

An older man stepped forward, his demeanor imperious as he introduced himself. "I don't know if you have heard of me, but I'm Noel Knowall." It was obvious that he had decided he should speak for his team, since he had been on previous hunts with his senior team leader, and clearly regarded himself as the most qualified to present his team's approach.

Jason resisted the urge to roll his eyes. He had never been on a hunt with Noel, but knew his reputation. "Yes, Noel, I have, indeed, heard of you. How are you doing?"

Noel smiled in his uniquely patronizing manner, which Jason chose to ignore. "Fine, sir. If I may answer that last question, over the past twenty years that I've been going on safaris, we have always taken the chickens with us. They're very useful, and, in my personal opinion, they'll help distract the crocodiles. We used the chickens with great success before on the grassy plains."

"Ever been on a safari in a swamp before, Noel?" Jason asked.

"Mr. Hunter, I've been on every other kind of hunt. We've been in the jungle, which is quite similar. Of course, we've spent quite some time on the grassy plains and in the woods. In my estimation, the same techniques and principles that have always been successful on other safaris should apply to this one, as well." Even though Noel had clearly not been on a safari through a treacherous swamp, his response was one of absolute confidence, and no small amount of arrogance.

Jason chose his words carefully. "I think we're in for some new challenges, Noel," he said. "More water, for one. These crocs are a new twist as well, but let's see how these ox carts and chickens do. I'll evaluate the SUVs as we go along." Jason was biting his tongue by now. He didn't like Noel's snooty attitude, and suspected he would be a problem before all was said and done, but he was determined that things would at least begin smoothly.

"All right, then, let's not get bogged down in analysis," Jason continued. "We created a plan, and we need to follow it. That brings me to another issue. I'm concerned about team organization."

"Jason," the Senior Safari Lead protested, "we made our hiring decisions based on the plan."

Trying to stay calm, Jason replied, "The teams are unbalanced. We need people to guide the safari and others to do the work. Some teams have too many managers and others too few. With too many workers, it's too easy to go off in the wrong direction. And with too many managers, nothing actually gets done."

"I'm afraid it's too late to restructure our teams now, sir," the Senior Safari Lead continued. "Everyone is here, has been briefed on their responsibilities, and we've pretty well committed to the plan."

Jason knew that his Senior Lead was right. For a moment, he considered re-organizing the team anyway, but thought better of it. He knew that morale would take a big hit if he undermined the authority of his Team Leaders at the beginning of the hunt, and that could prove more than a little dangerous.

The final pre-expedition confirmation continued. All the items in the checklist were accounted for, but some of the requirements were met in unexpected and indirect ways. The next week was spent staging the order of the march and relocating the base camp closer to the small village near the dirt road that led into the swamp.

The Hunt Begins

Water was plentiful in the small village, but some of the food and supplies that were supposed to have been there were not. Part of the team went back to negotiate for replacements in Supply City. The safari was delayed by a week, but they were only two days behind the overall schedule. Jason had experienced worse starts, and was not concerned – not yet, anyway.

With the expedition finally underway, the teams found the road to the swamp easy to traverse. It was a well-paved road, rather than the dirt road they had predicted. Jason, however, knew that the road would change as the safari progressed deeper into the wild. He'd seen it too many times before.

During this first leg of the journey, they made better progress than they had expected. In fact, they now were less than one day behind schedule. Jason began to relax. The teams were marched in a particular order according to who would be needed first. This put the ox carts in the middle and the SUVs at the end of the line. Needless to say, the Safari Lead with the SUVs was frustrated, impatient at his progress being impeded by the antiquated and painfully slow vehicles at the front.

At the entrance to the swamp, the group encountered several simple wooden roadblocks, which they easily moved aside. No one paid heed to their warnings of danger ahead. The leads all knew, of course, that there might be some crocodiles in the swamp, but they knew that danger and risk were a part of any hunt. They had made a list of known risks and felt prepared for any situation that should arise. None of the risks were seen as threats to the success of the hunt, and everyone felt confident and upbeat.

As they entered the swamp, however, some of the younger hunters expressed surprise at the large number of crocs in the water. No one had told them there would be so many, only that the crocs would be there. The reality that they now encountered made the younger hunters uneasy. Their uneasiness was amplified by their limited perspective and relative inexperience. None of them knew much about this safari beyond their own jobs, and most of them had only the most general picture of The Safari Plan; furthermore, they were used to open plains and marshes, not overgrown swamps teeming with crocs.

Stomp Yer Croc!

Despite the large number of crocs, however, not everyone saw them at first. The nasty critters were well camouflaged and kept their distance, so it was some time before everyone really became aware of them. Even then, some of the more senior members of the team claimed the sightings were exaggerated, and that any crocs that might be there posed no problem. "It's all part of the challenge of a hunt!" huffed one Senior Hunter. "Besides, all of the maps place the crocs much deeper into the swamp area than we are now."

These patronizing reassurances placated some of the young hunters for a while, but didn't entirely erase their anxiety. Those crocs might not seem like an immediate threat, but who knew how many were really hidden in the murky depths of this mostly unknown territory? They suspected that the senior hunters were blind to the very real danger they all faced.

"It's all part of the challenge of a hunt!"
the Senior Hunter smugly reassured them.

Facing The Swamp

It took two days to get the entire safari onto the swamp road. Actually, calling it a road was too generous. In truth, it was little more than a trail, narrow and choked with vines. Cutting through the undergrowth took time and slowed their progress.

And then there were the monkeys. Ever since they'd entered the swamp, the safari members had noticed large gatherings of monkeys up in the trees. As the teams cut their way through the swamp, the curious creatures frantically swooped down out of the trees, eager to check out these strangers in their home. The monkeys were playful and surprisingly friendly, and most of the expedition members found them quite amusing, but they slowed the safari's advance. Even Noel got distracted by them, and actually made a pet of one. When the monkeys started eating some of their supplies, however, one Safari Lead got angry, and admonished the others, saying, "Stop spending time with those monkeys! They're a nuisance."

To which Noel replied, "Oh come on, man! They're fun. Especially this little fellow here. I've named him Minkee the Monkey. Right clever, don't you think?"

"No, I don't, Noel," the Safari Lead replied. "He and his friends are more hindrance than help. Those damnable pests are a complete waste of time, and they're eating our supplies!"

Seeing that his warnings weren't so much as making a dent, the Safari Lead stomped off in disgust. Noel wasn't on his team, anyway. Noel returned to the march, with Minkee perched happily on his shoulder. Whenever the safari stopped to rest, Noel amused himself by playing with Minkee, and it soon became obvious that Noel wasn't fully attending to his duties as a Team Lead. Between tending to the chickens and playing with Minkee, Noel didn't get much else done.

By the end of the first week, everyone was seeing crocs. They remained off in the distance, however, and showed little interest in bothering anyone. The team leaders and Safari Leads agreed that they would take care of them when they really became a problem. No one mentioned the crocs to Jason. He saw them, of course, but like his Safari Leads, he thought nothing of them, since they weren't actually hurting anyone.

Besides, there were other concerns. There wasn't any grass for the oxen to eat. In addition, the care of the chickens and the oxen was taking a lot of time, and detracted from the quality of their keepers' performance of their other camp duties.

Finally, Jason felt compelled to call a meeting with his Senior Safari Lead, who brought Noel with him. Jason was blunt but compassionate. "You and your people are overworked from caring for the oxen and the chickens," he began. "They're exhausted and so are you."

"It's a temporary challenge," the Senior Safari Lead replied. "We've overcome these kinds of inconveniences before. You'll see."

Noel shook his head and muttered, "We should have found a better road. The oxen are eating the grain we needed to make fresh biscuits for our teams."

"Do you have an alternative, Noel?" Jason asked.

"As a matter of fact, I do. There are passes through the mountains. All we need to do is travel up into the passes and rappel down the cliffs directly into the Valley of Gold. None of this swamp nonsense is necessary."

"Jeeps, SUVs, ox carts, and all?" Jason snapped. "How do you rappel them down the cliffs, much less get them out with the gold? Did you think this through at all?"

Noel didn't answer directly. "It's just that in this situation, we can't do things the way we've always done them."

Jason was beginning to lose patience with Noel. "What about those standard techniques you bragged about back in Supply City?"

Noel said quietly, "All right… of course we should use standardized techniques. They've worked countless times in the past. But I'm afraid that this dreadful trek calls for momentous modifications." The Senior Safari Lead stayed out of the argument. Jason, however, had grown tired of the growing problems with the supposedly perfect plan.

"I agree that we need to modify our plan, but innovation just for the sake of it isn't going to serve us," he said to Noel. "It has to be practical. We already know that the swamp road is the only way in. I need solutions that deal with where we are now."

Two nights later, the oxen started to disappear. The Safari Leads all thought that it might be the crocs' doing, but no one had any proof. After that, some carts had to be tossed into the swamp, since there were no oxen to pull them. The remaining oxen were carefully guarded in the middle of the camp.

By the third week, the younger hunters' anxiety – and their complaints – had increased, and they began shooting at the crocs, which would disappear in the murky water, only to re-appear elsewhere. The Safari Leads finally began to listen to their concerns, and they too began complaining among themselves, but didn't take any decisive action. Instead, the Safari Leads simply tried to motivate everyone to do a better job. Although they had to acknowledge that the crocs were everywhere, they still ignored the younger hunters' insistence that the crocs were seriously impeding the safari's progress.

The fuel supply for the Jeeps and SUVs began to run so low that the decision was made to leave half of them behind. The armored personnel carrier, however, was kept to transport the chickens.

After the team had been marching through the swamp for weeks, there finally came a point when the crocs presented a danger that was clear to everyone. Perhaps it had something to do with the fact that some of the beasts were now crawling out of the water and making their way to the trail in broad daylight, clearly unintimidated by the human intruders. The younger team members were stomping on the small crocs that they saw on the trail – satisfying, perhaps, but not particularly effective in solving the group's problem, as the large crocs grew increasingly brazen, devouring some of the remaining oxen in plain sight of everyone. One of the larger crocs actually smashed into one of the Jeeps, tipping it over into the swamp.

The problem with the crocs had become too obvious to ignore. The safari was two-thirds of the way to the Valley of Gold, but now found itself almost two full weeks behind schedule. All of the Safari Leads recognized that the expedition was in serious trouble. No one – not even Jason – could deny it any more.

Towards the end of a particularly trying day, they were preparing to set up camp at a large clearing in the swamp, a spot that had been marked on their maps. But as they approached the clearing they saw that it was surrounded by barriers of pointed sticks that blocked the entrance to the spot.

Noel, upon seeing the barriers, exclaimed, "Well, will you look at that! A C*hevaux-de-frise*, a type of defensive obstacle that's been around at least since the Middle Ages… I've seen them on hunts before. Excellent defense for all kinds of varmints, even crocs!" He looked around at the group with a smug and knowing expression on his face.

"I know what they are," Jason answered a bit sarcastically, glaring at Noel.

Only slightly chastened, Noel said, "I was just explaining things for the benefit of our younger friends here who might not know."

"Well, I am sure we all appreciate the history lesson," Jason said. "But we need to get these barriers out of the way so we can get to our camp site."

They all set to work. It took a little time, but the barriers across the trail were moved out of the way so the safari could enter the clearing. After a couple of hours' effort, the group began to set up camp, re-setting the barriers as a defense against the crocs. With the defenses around them, Jason agreed that it was time to re-group, analyze their progress, and quite possibly, re-evaluate their carefully laid Safari Plan.

Noel exclaimed, "It's chevaux-de-frise. Seen them on hunts before. Excellent defense for all kinds of varmints, even crocs!"

So the first thing Jason did was call a meeting. As the team was stalled in the swamp and in dire need of a revised plan, the meeting went on for three days. They re-planned, and then re-planned again. There were no best solutions – only compromises. Many of the Safari Leads and their most trusted staff argued that only win-win options were acceptable, even though it was increasingly obvious that very few, if any, such solutions were available.

The final plan they came up with left the oxen, the SUVs, and the armored personnel carrier behind. Furthermore, only a third of the Jeeps could make the entire trip, because even their most conservative fuel use estimates made it clear that there wasn't enough fuel to power all of them.

They kept the chickens, but, with the personnel carrier left behind, each staff member was assigned the task of carrying a chicken in a cage, along with his or her own food rations. The task would have been more difficult, and their burdens greater, save for the fact that all of the grain was gone, and since the cook stoves were out of fuel and thereby useless, they could be left behind, as well.

As the Safari Leads explained the new plan to their teams, they tried to put the best possible spin on it. The younger hunters were happy that someone finally was listening to them and acknowledging that the problems they had foreseen were already emerging. The more experienced team members, however, were not reassured. They sensed that they weren't being told the entire story, and began to voice their own concerns. It didn't take long for their dissatisfaction to spread to the younger hunters and dispel their brief spurt of confidence.

Knowing that resistance to change is a common problem in any group – especially one as immersed in a stressful situation as theirs – the Safari Leads did their best to pacify those who complained the loudest. They reverted to their favorite mantra about rising to meet challenges. Their teams, however, saw real problems that The Plan utterly failed to address, and that were being summarily swept aside. They knew full well that these were no mere "challenges," to be conquered by a combination of platitudes and stiff upper lips. These were problems that could not only compromise the success of the safari; they could put their very lives at risk.

The Valley Of Gold

The expedition was a month overdue when it finally reached the entrance to the Valley of Gold. For a few brief moments, and especially among some of the younger members, there was a sense of celebration, of having finally reached the goal. A few of the team members exchanged high-fives. "We made it!" they said.

Jason knew better, and to him the "victory" was bittersweet. Having been on many hunts before, and knowing the cost of completing the return trip, he knew that this safari was badly over budget. Too much had gone desperately wrong. Their perfect plan too often lacked the right details. Much of the information upon which the plan had been devised had turned out to be either incomplete or totally wrong.

After all of the glitches and near-disasters, they did find some of the gold they sought, and a small amount of silver, as well. At least these would pay for the trip, and there would even be enough to show a meager profit.

We did get through it without losing any of our team members, and we did get a little bit of gold, Jason thought to himself, *but was it worth the price or the effort? Besides, we've only just arrived at the entrance to this valley, and I know there's lots more gold for the taking if only we could venture further in.* He realized, however, and with no small amount of frustration, that he simply lacked the resources to go any further.

There was no sense torturing himself about what might have been. Still, he couldn't resist taking a closer look at the legendary Valley Of Gold before he left. And as he looked across the expanse of the valley, he was stunned by what he saw. It hadn't been apparent on first glance, but now before him, for seemingly endless miles, lay the entire valley, the very ground shimmering, laden with more gold than he could have ever imagined. Then, in the distant mists, he thought he saw movement among the grasses and reeds. Raising his binoculars, he focused on the broad expanse of treasure in the distance, and saw what had caught his attention: amid the piles of gold were literally hundreds of snarling and snapping crocs, standing guard like a legion of mythical dragons over their lair. He now fully realized that there was a fortune beyond imagining, lying just beyond his grasp, that he could have reaped, had the hunt only gone better.

"We failed," Jason said aloud. "We really failed. So much more is there, and we can't even take it."

As Jason continued to stare in amazement at the unclaimed gold, Noel spoke, interrupting his reflections. "Mr. Hunter, the expedition was a success!"

Stomp Yer Croc!

Jason, jarred from his thoughts, spun to face Noel, barely able to contain his anger. "A success? Hardly. We covered our costs and made a little bit extra. Don't you see how much more gold we could have taken out if we had actually gotten into the valley itself?" In that moment Noel, standing there and petting that stupid monkey of his, seemed to represent everything that had gone wrong with the expedition. It took a real effort for Jason not to say as much, but he knew it would serve no purpose, and wasn't the truth, at any rate. Jason turned away, looking deep into the valley.

Noel started to answer but decided against it. He thought to himself, *We always leave some gold behind. That's part of any safari. We never have done things any differently, and we never can. So what's the big deal?*

And then, finally noticing the piles of gold in the distance, Noel understood Jason's point at last. The two began walking toward the treasure closest to them. In the meantime, the rest of the team had gathered around, and now they too saw the piles of gold, as well as the largest and most menacing crocs they'd seen on the entire journey. The beasts really did seem to be standing guard. Behind these crocs, nests rose from the water. As the hunters all stood and watched in amazement, countless baby crocs hatched in the nests, and like their larger parents, seemed to begin hoarding the stash of treasure that surrounded them.

Jason had to acknowledge to himself that his former skepticism about the crocodiles had been somewhat modified. "I admit some of those crocs are pretty large, and they do seem to think they have some sort of claim on that gold," he said aloud, "but, as we can see for ourselves, there's apparently no such thing as King Croc."

The Safari Leads all agreed. "Just another one of those out-of-proportion myths," the Senior Safari Lead added.

Then a younger hunter pointed to one particularly huge reptile and said, "I wouldn't be so sure. Now *that's* a lot of croc." He began to stammer, "That's h-h-him…th-that's old K-k-k-king C-Cris Croc!"

"Now that's a lot of croc," exclaimed a younger hunter.

"I thought it was just 'King Croc'," a Safari Lead said.

The younger hunter replied, his voice still trembling, "Just to be proper, I thought he needed a first name." A couple of the other hunters rolled their eyes, but said nothing.

"Well, proper or not, that old boy is guarding at least a third of the gold. Perhaps more," the Safari Lead replied.

Disregarding the other hunters' observations, Jason shook his head. "I still have trouble believing in such nonsense. He's just another crocodile, no different from the others, except bigger and probably a bit meaner. Next time through – and there will be a next time, mark my words – we will be prepared. We'll have the people, the time, and the money we need to capture more of that gold. And the silver, too."

As the safari left the treasure-laden valley, Jason continued to think about lost opportunities. "How much more profitable could this journey have been?" Jason asked aloud. "Because of our poor planning, those crocs are keeping what could have been ours!"

It was a thought that would not leave him alone. Crocs or not, he knew he would be returning soon to the Valley of Gold.

Stomp Yer Croc!

two
The Nature of the Hunt

The hunt was finished. The group traveled in virtual silence along the swamp road, each consumed with his own thoughts of the expedition now behind them. It was nighttime when Jason finally left, however. There were too many questions in his mind, too much that he needed to consider without the distraction of the other safari members.

Climbing into a Jeep, Jason drove away from the base camp. He stopped on a hill overlooking the road into the Valley of Gold. "How can the next trip into the valley be more successful?" Jason asked aloud.

He had many questions about how to improve their next journey. Yet, each question suggested answers, and each answer suggested new questions. *Did we really have the right people, or were they asked to do things that didn't make sense to them?* he asked himself. *Some things did go wrong, but what went right?*

The sky was clear and filled with a million glittering stars, but he was so deep in thought that he scarcely noticed. *The next trip must be better planned and executed. What could we have done differently… and better? Who can help me?* As he mentally reviewed a list of names, another well-known and successful hunter, Alexander Parmenson, came to mind.

Alex had helped new hunters as well as those with years of experience, and he had the best reputation of any hunter that Jason knew or had heard of. *But will Alex work with us?* he wondered.

Jason climbed back into the Jeep and returned to the camp to gather his gear. When he finally left, it only took him a few hours to catch up to the others, who had already set their camp for the night.

The sky was clear. The stars shone brightly...
"What could we have done differently... and better?"

After several days, they returned safely to the base camp at Supply City, much to the surprise of the local villagers, who had felt sure that the journey would come to a bad end. They were more than a little surprised that everyone in Jason's expedition had returned, and they immediately asked about King Croc.

"Some of my people claimed to see him," Jason replied, "but I still doubt that he exists. I saw one large croc in the distance, but there was nothing special about him. He didn't seem that big. I figure that story has been around for many years, and if he ever existed, he is most probably dead by now."

Of course, the villagers assumed that Jason was wrong, and that he had either not seen King Croc, or had refused to acknowledge just how large a beast the croc actually was.

In truth, the latter was more accurate. Whether because of the distance from which Jason saw the croc, or simply the fact that he did not want to accept that the legends might actually be true, his perspective was a bit distorted. He had, indeed, seen the creature, and had grossly underestimated King Croc's size.

Once he had settled into his accommodations in Supply City, Jason called the bankers and financiers and gave them an account of the venture. He hadn't really been looking forward to reporting such a limited success, but to his surprise, they were, to a person, very pleased. "You did it again!" he heard over and over. With some reluctance Jason accepted their congratulations, but he couldn't really take their words to heart. His belief that the limited success wasn't worth the time and effort that everyone had put into the hunt continued to gnaw at him.

He spent most of the night in deep thought, but he wasn't merely beating himself up for his perceived failures. By the next morning he was eager to start planning a more successful second hunt. The first course of action was to sift through his contact information until he found Alex's phone number. It was still too early to call, what with the different time zones, so he waited, albeit impatiently.

When he finally decided that it was late enough to call Alex without waking him, he did so, only to get Alex's voicemail. He hated having to wait, but he left an excited message for Alex, and then fidgeted and paced for a few hours until his cell phone finally rang, with Alex on the other end.

Whose Point Of View?

"What a surprise to hear from you, Jason," Alex began. "Word of your success getting into – and out of – the Valley of Gold is the main topic of conversation at every power lunch in town."

Jason hesitated, taken aback by Alex's words. He couldn't help but wonder if he might be overreacting in his dismal assessment of the safari's success.

"Jason, are you there?" Alex asked.

"Yes, yes, I'm here. Alex, thank you for your kind words, but the hunt wasn't the success people apparently think it was. We had all kinds of problems. Besides, there's a lot more gold to be had, and we only brought back a fraction of what we could have. Actually, we only made it to the entrance of the valley."

"Ah, yes, public perception. Even so, making it in and out of that valley was an achievement in itself. And after all, you did make a profit. People see that as success. But you obviously don't. So what do you think went wrong?"

Again, Jason hesitated. "That's the problem. I'm not exactly sure. We had hot spots. That much I know. I think I need your help in analyzing this hunt, and if you're interested, in getting ready for our next trip to the valley."

"Hot spots? What kind of hot spots?"

"Just about everything," Jason replied. He had already been over this many times during his personal reflections the night before, so it wasn't all that difficult to express his thoughts to Alex now. "People, planning, technology, research, communication, and even the techniques we used. I'd really like some fresh eyes to look at what went wrong. And I'd like you to help me plan the next hunt."

"How soon will you be back in your office?" Alex asked.

"I've decided to give the team a week off. Whether they stay here or go home is up to them, but everyone is to be back here in Supply City a week from this coming Monday. So what do you say? Are you up for an… er… adventure?"

"Sounds like something I could sink my teeth into, all right. You get some rest over the weekend. I'll clear my calendar and make my travel arrangements. I'll be in Supply City this coming Tuesday. We can talk when I arrive."

Jason hated to wait, but knew that the trip to Supply City would take Alex at least three days. And if Alex was interested enough that he was willing to travel on the weekend, Jason figured the unavoidable delay would actually be a good thing. It would give Jason more time to really sort out what had gone wrong. "Okay, Alex, Tuesday it is."

Jason And Alex Meet

It was late Tuesday afternoon before Alex arrived in Supply City. Jason was anxious to talk, but gave Alex some time to settle into his quarters first. They enjoyed a casual dinner in a nearly empty headquarters tent. Some of the staff had declined the offer of a week off and had remained in Supply City. They all knew of Alex, whose reputation, like Jason's, preceded him.

Even Noel, who normally harbored little respect for anyone, greeted Alex warmly. "Are you going with us on the next hunt?" Noel asked with something approaching eagerness – a very un-Noel-like response.

"Not sure yet," Alex answered. "A lot depends on what Jason wants to do. This is a fascinating hunt and a difficult one."

"Which brings us to the very reason I invited Alex to join us," Jason quickly piped in, not wanting to spend any more time with Noel. "Alex, are you up to a short meeting in my tent, or would you like to rest up a little longer?"

Smiling, Alex replied, "Sure. After all, it's not like real work. We'll be talking about our favorite activity – treasure hunting."

As the day waned, the team members departed for their respective tents. The sun had set by the time Alex and Jason sat down at the table in Jason's tent. "Okay, Jason," Alex began, "let's really talk about what went wrong."

"Like I said when I called," Jason replied, "there were problems everywhere. I thought we had a perfect plan – the best I'd ever seen. I really believed we had every aspect of the safari covered." Alex didn't respond. He just looked at Jason, waiting for him to continue.

Stomp Yer Croc!

Jason stared off into the distance for a moment, then went on. "I spent a lot of time putting a comprehensive plan together. I selected the best Safari Leads I had or could find. All of them are experienced. I thoroughly reviewed the overall plan with each of them. Once I felt comfortable that they understood what was expected, the Leads spent several days putting together their own plans and hiring their own teams."

He continued, "Because the Safari Leads spent all that time planning as well, I thought they had every angle covered. That's the way it's supposed to be. But in spite of these efforts, during the safari, none of the Safari Leads seemed to be following their own plans, much less doing anything even remotely in sync with mine."

"Can you give me an example?" Alex asked, as he sipped his coffee.

"Sure," Jason replied. "We agreed on the need for appropriate transportation for the hunt. The Safari Leads brought everything from ox carts, Jeeps, and SUVs to an armored personnel carrier. They were all different!"

Alex chuckled. "Ox carts, you say. Do these folks still use typewriters, or haven't they given up their quill pens yet? It reminds me of my first safari. You did leave the ox carts behind, I hope."

Obviously embarrassed, Jason looked down, mumbling, "No. I let them use their ox carts."

"Let's start with some basics." Alex said, hoping to change the subject and let Jason off the hook a bit. "We'll talk to the whole team on Monday and chat about the problems they had, including transportation."

Jason was surprised that Alex wanted to meet with the entire team, especially so quickly. Jason usually preferred to meet with the Safari Leads and their team members independently, so as to avoid any friction and personality conflicts. He also felt he could get more honest answers about any problems that were brewing.

Then Alex clarified his intentions. "Before we meet with the whole team, we'll talk to the Safari Leads alone." Jason felt a surge of relief. This was more in keeping with what he was accustomed to doing.

Alex stood up. "For now, though, I'd like you to draw a five-pointed star," he said, pointing to the clipboard in the corner of the tent.

"We have some planning to do to get you ready for those team meetings on Monday."

Alex slid a sheet of paper toward Jason, who grabbed a marker and drew a large star. "Okay, write the words that I tell you. We'll look at one starpoint at a time," Alex began. "We'll name the starpoints now, but you'll have several questions to answer for each one. Start on the point on the upper left and go clockwise.

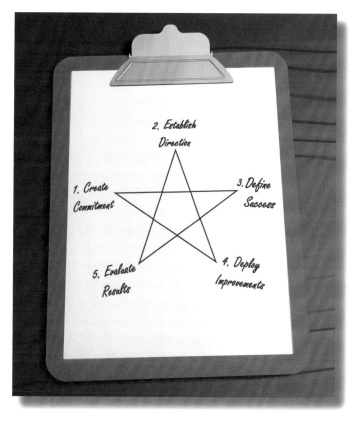

- The first point is to **Create Commitment**.
- Second is to **Establish Direction**.
- Third, **Define Success**.
- Fourth, **Deploy Improvements**.
- And the fifth and last point is **Evaluate Results**."

As Alex spoke, Jason wrote. It only took a couple of minutes, but Jason began to see the familiar high-level concepts of the five starpoints.

Stomp Yer Croc!

"Looks pretty simple to me," Jason said. "I did all of these things. Seems to me we're doing number five right now."

"We are, and we'll do it by looking at each starpoint. But doing a plan just to say you planned isn't enough. Your plan must be useful and not merely done for the sake of planning. And you know very well that you don't have to accept everything a team member suggests; you just need to make them feel they're involved."

Jason looked at Alex in amazement. "What do you mean? We planned every stage of the hunt in de…." He stopped himself, realizing what he had said. "We're talking about the ox carts, aren't we?"

"My guess," Alex said, with a gentle nod and a smile, "is that we're going to be looking at some other details, too. For the moment, let's plan for the meeting with the Safari Leads. Your Safari Leads can give you answers. Then, we can meet with the whole team."

A Shaky Team Meeting

Alex and Jason spent the next three days getting ready for the initial meeting with the Safari Leads. They also talked in great detail about commitment, communication, and organization. Alex spent some time on his own as well, talking with the local villagers.

After one of his forays into the village, Alex asked Jason a surprising question. "What about King Croc?"

"He doesn't exist," Jason replied with a smile. "Villagers got you with that fairy tale, didn't they?"

"Why do you think he's a fairy tale? The villagers say your people saw him."

Jason reacted with unexpected emotion. "Oh, they saw one slightly larger croc, and I saw him, too. He was near a large stash of gold. With that I'll agree, but this 'King Croc' business is just going too far."

"'King Croc' or a large croc, he's a risk. If you want more gold and silver, then you need to prepare to deal with him," Alex said firmly. "No reasonable concern or risk can be taken off the table."

Over the weekend the two men relaxed, visited the small town near the swamp, and fine-tuned their approach. Alex had met and talked with most of the Safari Leads individually the previous week, and spoke with the remaining two on Saturday. These meetings gave Alex some insights into their concerns. When Monday morning came, Jason thought he was ready for the meeting. And the Safari Leads, after their interviews with Alex, were certain they were prepared as well.

The first round of meetings, with only Alex, Jason and the Safari Leads, took place in the main headquarters tent. Food was brought in so that they could work without interruptions. Jason opened the meeting and formally introduced Alex as a new member of the team. Then he opened a flip chart with the meeting agenda on the first page.

"We've all been introduced," Jason began. "Alex has talked with each of us. Now, we can get comfortable and get down to work. I'll let Alex run our meeting, including the 'Lessons Learned' item. He didn't have a dog in this hunt, so he'll be more objective."

Alex tore the agenda off the flip chart and attached it to a tent pole. Picking up a marker and drawing a line down the middle of the clean sheet, he wrote, "What Worked" on the left, and "What Didn't Work" on the right. "All right," Alex began. "One at a time, let's have at it."

Jason's Senior Safari Lead spoke first. "The personnel carrier was a waste of time and money. It lumbered along and used more fuel than three Jeeps."

Alex started to write 'personnel carrier' under "What Didn't Work," when the Safari Lead who had brought the personnel carrier shot back, "And your stupid ox carts slowed everyone down!"

"Well, you hired all senior hunters, and my people ended up doing all the work," the Senior Lead retorted.

Alex knew this group was bitter about the hunt, and he had to stop the bickering. "Guys, let's not get personal."

"He started it," the Safari Lead with the personnel carrier said.

"That's not the point," Alex interjected. "We need to solve problems, not make more of them by launching personal attacks. Neither the ox carts nor the personnel carriers worked. We need to know why."

Jason was smiling. "No one's mentioned those chunky old chickens yet. In my opinion, we should have fed them to the crocs."

"The personal attacks go for you, too, Jason," Alex replied. "The chickens had some use. They gave you fresh eggs for one thing, and meat later in the hunt. Of course, there are more efficient ways to get both."

"Sorry, Alex, it's just that there's so much to be fixed."

Alex stopped writing on the flip chart and said, "I think it might be a good idea for us to stop right now and go over some ground rules. First off, everyone here was included on the safari because they had specific skills, knowledge, and experience that were needed. None were selected because they were perfect. We need to keep these facts in mind if we're going to actually solve the problems the expedition encountered. Remember that those problems arose in spite of all your expertise. We need to look at the what and the why, and forget about the who. Each of you deserves to be shown equal respect. So I ask that all of you listen, and make no statements attacking other people."

The team – including Jason – was still smarting from the failure, and most of them were reluctant to let go of their frustration at first. But after a bit more back-and-forth, they understood what Alex was saying and agreed to the new rules. "Okay, let's start again. What worked? What didn't work?"

"The SUVs were too hi-tech for the swamp," the Senior Lead added. "They could go fast, but the Jeeps worked better. Also, the GPS failed because we lost the signal once we were deep in the swamp."

Alex smiled. "Now, that's more like it." From that point on, the meeting progressed with less conflict and more cooperation.

Building On Experience

The team spent all day Monday and well into the evening hours working through the agenda. By late afternoon it was apparent that the meeting with the entire crew was going to have to be postponed until the next day.

For a while, Jason was a little annoyed about falling behind schedule, but he trusted Alex. He knew Alex had not gained his stellar reputation by wasting his own time or those of his clients. Surely all of this careful review, and these endless discussions, would help ensure a truly successful trek.

Their list of "Lessons Learned" was not only invaluable for planning the logistics of the next trip, but it also proved to be useful in shaping the safari's vision, mission, and core values. At first, a few members of the team were somewhat resistant to these ideas. Most had a passing familiarity with the concept of vision and mission statements, and those who had worked with Jason were vaguely aware that he had some sort of mission statement for his company.

But some of them didn't think it was necessary to bother with seemingly abstract notions now, especially since there were so many concrete issues still to be dealt with. When Alex turned to a clean page in the flip chart and wrote the words, "Vision and Mission," one of the Safari Leads began muttering to the man sitting beside him.

"Dan, do you have something to say about this?" Jason asked the Lead. To tell the truth, he was a little puzzled himself about why Alex wanted to spend time talking about these things, especially since their "mission" seemed pretty clear to everyone there.

Dan looked a bit embarrassed, but cleared his throat and said, "I was just wondering if it's really necessary to be spending so much time on high-falutin' statements about what we intend to do and how we're going to change the world and all that… when all we really want to do is grab all the gold we didn't get the last time because of the mistakes we made."

A couple of other Leads murmured agreement, and one said, "Yeah, I mean, most of us read 'Dilbert.' We're not a bunch of suits who spend our days drinking coffee around a conference table and dreaming up new ways to waste time. We just want to get the job done." There were more murmurs of agreement, and everyone looked at Jason and Alex.

Jason was a bit uncomfortable, even though he secretly thought that Dan and the others had a point. He had hired a consultant a few years back to help him craft his vision and mission statements for his company, but that was mostly because it seemed to be the hot trend at the time. Even so, he had to admit he had drawn some strength and inspiration from those statements, and they had helped keep his company on course. He really didn't see the need for hammering out a vision or mission statement for this particular trek. He was unsure about what to say to the team now, and he felt the full weight of the momentary awkward silence that followed the Leads' comments.

To Jason's great relief, Alex smiled broadly. "Hey, I completely understand why you guys would feel this way," he said. "But rest assured we're not going to spend hours composing lofty nonsense that has no practical value for this hunt. I'm not here to waste your time, or mine. I want you to get the gold just as badly as you want it – the sooner the better! But as you learned on the last trip, it's not that simple. We've spent a lot of time so far going over the concrete details about what went wrong before.

"Now we need to dig a little deeper so we can really get it right this time. Agreeing that we want to get the gold just isn't enough; a good, solid vision and mission statement will help us clarify that goal. And I know from experience that coming up with a great list of core values will help us from going off course logistically or ethically."

That seemed to mollify the dissenters, and it put Jason's mind at ease as well. And so they set to work. Separating the vision for Jason's company from the mission of the safari was more difficult than expected. Many of the leads, who had been on numerous safaris with Jason, believed that the vision belonged exclusively to him, and was not open for discussion or challenge. Alex explained that their ownership of the vision was necessary for their belief in the safari's mission. That made sense to them. Ultimately the group agreed that a safari-specific mission was to be applied to each trip – whether to the Valley of Gold or anywhere else. Alex also spent a considerable amount of time discussing the group's core values, listing those that he felt the team would need in order to be successful.

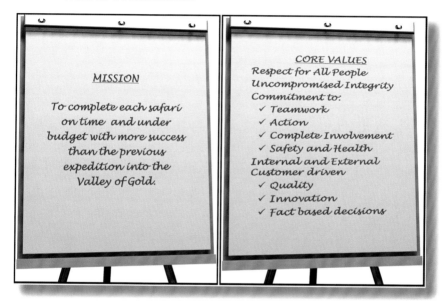

The vision, mission, and core values were completed by mid-afternoon. Only the issue of organization remained for them to consider.

At that point, the most junior Safari Lead interjected, "My team's organization worked well enough."

"Yes," Alex agreed, "and it may well be a model for the entire safari. Does everyone agree that his team's structure is a good starting point?"

As Alex expected, some dissent followed, mostly from other Team Leaders' hurt pride. They found, however, that the lessons learned during their discussions pointed to possible solutions. After the discussion had gone on for a while Alex determined that it was time to move on, rather than allow individual issues to dominate. "I think it's time for all of the Safari Leads to gather their team leads or senior hunters. We can brief them on what we've discussed. They may provide some insights that we haven't considered."

Stomp Yer Croc!

"Okay, let's meet again tomorrow morning," Jason said. "I know that we had originally planned to have the whole-team meeting today as well, but, as is often the case, this took a bit longer than we'd planned. That's all right, though. We covered a lot of ground, and I'm happy with the way it's going so far. We'll finish up our re-organization and then look at the Safari Plan." Though he was still unsure of the wisdom of bringing the senior hunters into the process so soon, Jason kept his concerns silent for the time being.

Involvement Means Commitment

After the meeting ended, some of the Safari Leads returned to their tents, while others went into Supply City to browse before dinner at the base camp. Alex, sensing that Jason had something on his mind, suggested that the two of them go for a walk, someplace where they could talk privately. They found a grove of trees where they would be protected from the sun, and they sat down.

"What's the problem?" Alex asked.

"Bringing in the senior hunters means that Noel will be part of the group," Jason said. "He's always giving me unwanted advice, and frankly, the man just irritates me, without even opening his mouth."

"I'm not going to pull any punches with you, Jason. You claim that respect for all people is one of your core values. Noel has value, or you would have sent him home a long time ago. Your Senior Safari Lead would have, too."

"They are a team, aren't they?" Jason responded.

"Yes, they are," Alex answered. "Noel makes up for what your Senior Safari Lead doesn't do well."

"Noel is going to be real upset about us not taking the ox carts or the chickens on the next hunt."

"They're familiar to him. Ox carts and chickens are his security blankets. He'll be even more upset about leaving Minkee behind," Alex chuckled. "I anticipate that he'll have a great deal of trouble adapting, and it's possible he simply won't be able to."

Jason was quiet and thoughtful for a moment. "Why not?" he asked.

"He's like a lot of people," Alex responded. "Change is very difficult. He also wants your respect. I know you don't believe that, but think about it. Every time he gives you unwanted advice, he's trying to build a relationship with you. All you need to do is give him the respect he needs. You do that by listening to and communicating with him to involve him in the process. You don't have to agree with everything he says or suggests. You know you can agree to disagree… agreeably."

Jason smiled.

"Besides," Alex added, "a lot of the team respects him – even the younger ones who find him a bit old-fashioned. They look to him for the guidance you aren't providing."

Jason was startled. "What do you mean?"

Alex turned around to look Jason squarely in the eyes. "You are committed without being 100% involved."

"I thought that I created commitment in others. I mean, I know I was committed to the hunt; I spent a lot of time planning it."

"Let's chat about that for a moment. You are the leader, but what does 'commitment' mean to you?"

"Well, I planned the hunt, I organized the hunt, and I found the money to get us there and back. I was involved in part of the planning process with each Lead. My reputation was on the line, and I'm always obligated to my partners, the bank, and the financiers," he said.

Alex leaned back and was silent for a moment before saying, "So you were committed to the bottom line."

"I'm the owner, so I'd better be committed to profit," Jason replied defensively. "If I wasn't, we wouldn't go on any more hunts."

"Did you attend any of the planning meetings held by your Safari Leads?"

"I didn't feel that I needed to go to any of them. My Senior Safari Lead went and reported on what happened."

"How about the transportation meeting?"

Jason didn't answer, but Alex already knew. "Did you read the reports on the Safari Lead meetings with their senior hunters?"

"I delegated authority like I should, Alex. I didn't want to micromanage. That's what I am paying my Senior Safari Lead to do."

"If you'd been more involved or, should I say, committed, you'd have known that attending those meetings would have been the right course of action. As it was, you missed getting critical information from your Safari Leads. Your key people, including you, must know what went on in every meeting involving significant decisions."

"Or I miss a critical detail. Is that right?" Jason asked.

"Give that man a cigar! Unfortunately, that's only part of commitment," Alex said. "I'm not saying that you should micromanage or go to all meetings. But you do need to communicate – and to be open to communication. When your people know where you are leading them, they'll perform better."

"Let's go back for dinner. I think I need to spend some time with Noel," Jason said.

He's learning, Alex thought to himself. *We're going to make real progress now that he understands involved commitment.*

A New, Better Plan

Tuesday morning, the entire team got together at last. The morning was spent building a new organization, based on more clearly defined roles and responsibilities. The new structure was more balanced, because this time, it was based on what each team needed to do on the hunt. Some teams grew, while others shrunk. Just before lunch, Alex had them set the discussion of the organization aside.

During lunch, they started applying the "Lessons Learned" to the old Safari Plan. Huge pieces of the former plan were completely gutted. The team also began looking at the data from the first hunt. The daily fuel consumption and safari progress data led to major revisions in both the schedule and budget. Jason put in place a reporting system so that he could monitor key events at the end of each day.

When discussing the shortcomings of the previous trek, the participants focused on the problems and not on pointing fingers of blame. "During the last hunt, many of the new hunters didn't know how to use their machetes or how to keep them sharp," a senior hunter observed. "We need to focus on training these folks."

Alex agreed. "They probably slowed your progress."

The senior hunter nodded. "That's not all. Many of us didn't learn until later in the hunt that there are ways of conserving the cooking stove fuel. When we ran out, we had to use campfires, and dry wood was in short supply."

Noel added, "Drying wood out takes time and lowers the cooking temperatures. Most nights it was dark before we'd finished cleaning up after the evening meal."

"Let's bring 10 percent more cooking fuel than we think we need," Jason offered as a solution. "If each of us carries a little bit extra, it won't add too much weight. And we also have to remember that part of the problem on the last trip was that we were behind schedule, which meant we needed more fuel than our original estimate." Everyone nodded in agreement.

They also decided to put small trailers on the Jeeps to carry some of the heavier equipment. Noel was delighted. "It's just a mechanized ox cart!" he declared.

Stomp Yer Croc!

"It's just a mechanized ox cart!"

"Overall, I think we're more ready this time," Jason said at last. "There are fewer things to slow us down."

Alex wasn't entirely convinced. "You aren't prepared for King Croc or, as your team calls him, Old Cris Croc."

"Alex, I was *there*. I didn't see King Croc, Cris Croc, or any other larger-than-life croc. There were just some bigger crocs." He looked around and saw that a couple of the hunters had expressions on their faces indicating that they didn't agree. But they said nothing, and he turned away from them; he didn't want to get into another discussion about mythological creatures.

"Call them what you will, Jason," said Alex, "are you prepared for them?"

Jason wasn't about to admit he was wrong. His team had sworn that they'd seen one really huge croc guarding a very large stash of gold, and he too had seen an enormous croc, but they all looked as if they could be overcome. "Yes, Alex, I think we can handle them."

As Alex walked back to his tent he thought, *That big croc is there. When it comes at Jason, he won't be ready.* The thought lingered in his mind for some time, and it was well past midnight before he was able to fall into a restless sleep.

Stomp Yer Croc!

A Fresh Hunt

three

The order of march into the swamp was the same as it had been for the first hunt. Those whose skills would be needed first took the lead, accompanied by Alex and Jason's Jeep. Jason was prudently cautious, but he could not help feeling confident that this hunt would be more successful than the last. After all, he reasoned, they had planned more carefully, eliminated the SNAFUs from the last hunt, and were prepared to measure their progress on a daily basis.

For his part, Alex was excited about being in the field again and working on a real expedition. The past years of just being an advisor had left him feeling somewhat out of touch. Memories of past hunts flooded his mind. *How much different will it be this time than the last?* he thought to himself.

Alex observed the difference in Jason's attitude. His intensity was unchanged; however, he was more relaxed and open to talking with – and listening to – each team member. It wasn't just superficial chatter either. He spoke with them more as equals, understanding that they knew what was expected of them.

They exchanged some good-natured jokes about choices of communications. Most of the older hunters carried the old-style field phones, as well as walkie-talkies. The younger hunters brought their smaller cell phones with the walkie-talkie feature built in. Jason didn't object, as each team had both equipment types.

The younger hunters couldn't imagine why the older hunters would prefer the bulky antiques. They needled their older teammates, claiming that the extra weight would likely slow the "old" guys down. It was quite obvious that they considered their more modern cell phones to be superior in every way.

Stomp Yer Croc!

"Those young hunters may be in for a surprise," Alex murmured to Jason. "There's likely to be some 'I can't hear you now' going on if they can't find a signal source." Jason nodded in agreement.

Still, the trek was going smoothly thus far. "Everyone's happy again," Jason observed aloud as the safari moved out, "but then again, we aren't in the swamp yet!" He knew the biggest challenges still lay ahead.

As Jason and Alex pondered what lay ahead, the safari moved out.

Better Progress

Without their progress being impeded by the ox carts, the team needed much less time to reach the entrance to the swamp than on the previous journey. The men on foot easily kept pace with the Jeeps. And yet, as much as he tried to maintain his level of cautious optimism, Jason couldn't avoid a slight surge in his anxiety level once they reached the swamp. This was the point where all of their careful planning would really be put to the test.

Jason's first concern was whether or not slashing through the undergrowth would slow progress as it had on the first hunt, but his worries about that issue were allayed when he overheard two young hunters talking. "I'm glad they showed us how to use these machetes," the first one said.

His friend replied, "I'm cutting through faster, but then, the vines don't seem as thick as before, either."

"It's because we cut through this path on the first hunt. Of course, we now have sharper machetes, thanks to Noel's lesson," his friend added. Jason smiled to himself, glad that he had listened to Alex's advice about Noel.

Although they moved faster than on the first hunt for the first couple of days, their progress still remained slower than planned. Once the team hit its stride, however, everything fell into place, and by the fourth day, they were advancing at a much quicker pace.

All the while, Jason kept scanning the swamp for signs of crocs, but he didn't see them anywhere. Turning to Alex, he said, "We're deeper into the swamp now, and on the last trip, we'd already seen crocs by the time we reached this point. In fact, this is where quite a few crocs were sighted on the old satellite telemetry report. But I don't see any now. I wonder what happened."

Alex didn't reply. He'd seen a few crocs off in the distance.

"Alex, did you hear me?" Jason asked.

"Yes, Jason, I heard you," Alex said at last. "The crocs are there. They're just avoiding you for the moment; give them time. You have a plan. Just stick to it."

Sure enough, two nights later, some of the junior hunters saw yellow eyes glowing in the water. They reported them to their senior hunter, who spoke with their Safari Lead. No one shot at the crocs; nonetheless, the planned defenses were put in place.

As they got deeper into the swamp, more problems sprouted up. By the end of the first week, the cell phones had stopped working. Although everyone used the generator-based electricity to charge their phones, that wasn't the problem. As Alex and Jason had suspected, they'd lost their signal source.

Now it was the older hunters' turn to tease the younger ones. "See, we old guys sometimes know what we're doing. We've done this before, after all. We've learned from experience."

The younger hunters took the kidding in stride, with a few rolling their eyes as if to say, "Yeah, yeah, we've heard all of this before." But they had to admit, at least to themselves, that the "old guys" had a point.

At last the crew reached a fork in the road that had previously been discussed at an impromptu last-minute planning meeting Jason had held with one of his Safari Leads and a couple of the senior hunters before their departure. Alex had not been at that meeting. The maps indicated the possibility that one of the side roads at this area connected with another road leading to the clearing with the *chevaux-de-frise*. In the interests of trying to find a shortcut for use on future trips, the team had agreed to send one group to investigate this possible alternate route. Jason met now with the Safari Lead and his two senior hunters to discuss the details once more.

"Remember, guys, the satellite telemetry showed a triple canopy of vegetation covering three miles of the trail," Jason pointed out. "We don't know if that trail connects, or if there's deep water."

"I know, sir," the Safari Lead replied, "but as you told us, there's only one way to find out for sure, and that's to check it out for ourselves. And if it works out, we'll be at the clearing two days before the main column of the safari arrives."

Jason nodded. "You have nine days worth of supplies, and you're going to be on foot. Without the Jeeps, you still can make good time. Stay in touch for as long as you can. We'll wait for you at the big clearing, if you don't arrive before us."

"Will do, Jason. We have extra batteries for our walkie-talkies, wrapped up to keep out the moisture until we need them."

They shook hands, and Jason said, "Good luck and be careful."

"We will, Jason," the Safari Lead answered.

As the special detail proceeded down the road, Jason looked at Alex. "I really do hope they find that shortcut. It sure will help on the next hunt."

Alex was silent for a few moments before he spoke. "I've led expeditions in the area twice before," he replied, "and I've never taken that trail. From what I've observed, there's no evidence that it will give you anything useful."

Jason was truly surprised to hear that Alex had been in the region before, and wondered if he had ever attempted to enter the Valley of Gold. He made a mental note to ask Alex about it later. For the time being, however, Jason was beginning to see why Alex understood their problems so well. Even after considering Alex' latest advice, however, Jason's mind was made up. "Alex, it's just a minor risk. For us, it's low cost and possibly high benefit. For my money, it's worth trying."

Dealing With Monkeys

Just as the team was getting bored with the day-to-day grind of slogging through the swamp, the monkeys appeared. Most of the men and senior hunters simply ignored them, so focused were they on getting to the clearing to take a much-needed rest break.

Noel, however, had started to miss his pet monkey, Minkee. The other monkeys were of no interest to him, but Noel kept talking about Minkee with the members of his team. He was careful not to mention Minkee if Jason was nearby, but Jason heard about Noel's obsession anyway. And he knew that this obsession was almost as distracting to Noel's teammates as Minkee himself had been. After discussing the matter with Alex, Jason decided to talk with Noel alone, and not involve the Safari Lead. Jason asked Noel to walk with him.

"Minkee was enough of a distraction on the last hunt when he was here," Jason began, not bothering to hide his frustration. "Talking about him all of the time is almost as bad."

Jason's dressing down stung more than intended, and Noel's feelings were hurt. "Jason, I still think he could prove to be useful. He was a smart little guy; he could do all kinds of things! With a bit of training he could be a real asset. We just haven't given him a chance."

Stomp Yer Croc!

Noel couldn't help himself. He kept talking about Minkee.

As they spoke, Jason got even more impatient. He had grown fed up with the monkeys on the first trip, and was exasperated that he was still having to deal with them. Sure, monkeys were cute, and they could be entertaining. But as the team had seen for themselves on the previous venture, the monkeys could be annoying and downright nasty.

They made an ungodly noise at the slightest provocation. They seemed to get a thrill out of jumping out of nowhere and landing on the shoulders of unsuspecting team members, pulling the men's hair, or grabbing their hats. Worst of all, they helped themselves to any food that wasn't closely guarded, and wreaked havoc on any equipment that they got their hands on. And Jason had to admit that for those working with him, Noel was as getting to be as much of a distraction as the monkeys.

Focusing more on these problems now than on Noel's feelings, Jason shot back, "Minkee and his friends just wouldn't be helpful on this kind of hunt. I thought we'd pretty well established that on the last trip. They detract from our efforts instead of adding to them. Sure, they can be good for a laugh at times, but for the most part they're annoying as all get-out. I know you like Minkee and you think of him as a pet. I don't have a problem with that. But a business venture – especially one that requires the level of coordination that this one does – is not the proper place for a rambunctious pet. So please don't bring him up again – not to me, and not to the rest of the team. It's distracting."

Chastened, Noel nodded and then walked back to his tent. Jason felt a slight twinge of remorse for having been so blunt, but then again, how else was he going to get his point across? Noel apparently did get the point, and he stopped talking about Minkee. As a result, his team began working more effectively.

That was not, however, the end of the problems with the monkeys. A few days later, the safari's supplies started to disappear, and Jason knew what was happening to them. He discussed the problem with the Safari Leads and the senior hunters, and they came up with some new storage and security measures that they felt would effectively monkey-proof their supplies. Once they implemented their ideas, the supply losses diminished to a more manageable level.

Managing Small Croc Herds

The special detail tasked with forging down the alternate branch had stayed in contact with Jason for the first two days, but for the past three days, he hadn't heard from them at all. Now, a day before arriving at the clearing, he began to feel concerned.

Then, while Jason was deep in thought about what had befallen the other group, the main column suddenly encountered numerous crocs blocking their path. He immediately halted the expedition and met with his Safari Leads.

"We were bound to come across crocs eventually," the Senior Safari Lead said.

Another Lead added, "True enough, but what do we do about them?"

Alex intervened before Jason could respond. "We have the approach we decided on at Supply City. Find the nest, get rid of it, and the whole herd of crocs will go away."

"That sounds great," Jason said. "But where's the nest?"

"I think it must be close by. That's why they're here on the road," the Senior Lead offered.

"If memory serves, groups of crocs are called floats, not herds," Noel interjected.

"That's beside the point at the moment, Noel," Alex shot back. "Our problem is getting rid of them without becoming their dinner. If you find the nest, how to deal with the crocs should be obvious."

The Safari Leads and their teams spread out until one of the groups found the nest. One of the younger hunters said, "Let's destroy this nest and be done with it!"

Just as the hunters were unslinging their rifles, one of the Safari Leads, said, "Wait! We don't have to destroy the nest. We only need to move it!" The others looked at him questioningly as he continued, "The croc herd is here because the nest is near the road. We can't leave the road, but if we move the nest, they'll move away with it. If we destroy the nests, the crocs may attack with a vengeance, and we'll have created an even worse problem."

To everyone's amazement, moving the nest seemed to work. A few of the younger hunters acted as decoys, waving their arms and drawing off the croc herd. A couple of the older hunters carefully moved the nest with long pieces of bamboo, being careful not to touch any of the eggs or baby crocs. It took a little time, but their efforts were successful, and the croc herd moved away from the road to protect the nest, allowing the safari to pass safely.

Stomp Yer Croc!

"Thanks to our having followed our new plan, we've had to deal with fewer crocs this time. We're only a half day from the clearing, and even though we've seen some crocs in the distance, these are the first we've actually had to confront," Jason observed.

Alex said nothing, but once he had gotten him out of earshot of the others, he cautioned Jason about being over-confident. He reminded him that the trip into the valley was not quite two-thirds completed, and that he still had the return trip after that.

Jason nodded in agreement, but replied, "You have to admit, though, that this is so much better than our first go-round. The way we're doing things this time, I believe we can handle anything this jungle can throw at us."

"Let me remind you," Alex replied, "that your special detail is missing, and you still have the crocs in the valley to deal with. Don't get too cocky!"

Jason knew that Alex had a point. "I *am* worried about the special detail. If they had reached the clearing, we should've heard from them by now," Jason said. In spite of the problems that they had yet to encounter, however, it was hard for him not to feel self-assured. The teams were performing well, and overall they were making much better progress than they had on the first trip. Still, the special detail was missing, and he was anxious to reach the clearing.

They continued the march to the clearing in silence, the team members' thoughts filled with several things at once: the dangers before and behind them, the excitement at the prospect of the treasure they expected to find, and all the other random matters that occupy one's mind on a journey. Like the rest of them, Jason was lost in thought, his whole mind bent upon Alex's warning about the possibility of trouble ahead. He remained confident that the team was prepared for anything that might occur, but he knew better than to discount Alex's advice.

Emerging from his thoughts, Jason couldn't help but observe that the team members were cooperating in ways that would have been impossible on the first hunt. The younger and older hunters worked together to move the small croc herds, repelling the larger crocs whenever necessary. The monkeys were no longer a distraction, as even the young hunters avoided them. On occasion, Noel could be seen peering at the creatures as they danced through the branches above, but beyond his occasional barely audible grumbling, even Noel didn't seem to be paying the monkeys any heed.

Communications improved, because Jason wasn't just managing, he was leading, and the Safari Leads were ensuring that his message reached the youngest hunters. New ideas were offered by members of each of the teams, and this time, the Leads – and Jason – listened to what the members had to say.

Along with the usual measure of horseplay and good-natured kidding, many useful ideas came up at their nightly campfire meetings. Some of the ideas were so obviously valuable that they were acted upon immediately. Others would be tested for use on the next hunt. And still others, offered up purely for the amusement of the team, were met with laughter and applause. Seeing that not only had the team learned to address obstacles more effectively, but that they had actually begun to enjoy their efforts, Jason believed that his confidence was well justified.

Then, a quarter mile from the clearing, that confidence was challenged. The Senior Safari Lead reported that not just one but three crocodile nests sat squarely on the road, with a herd of crocs in front of each nest. Jason halted the column, turning to his Senior Safari Lead. "Okay," he said, "we knew this might happen. What do you suggest?"

"Well, we can't move these nests like we did the last time, because there are too many," the Senior Lead replied. "There's no way around it. We've got to stomp these crocs and eliminate the nests. It could get messy."

This wasn't the solution Jason had hoped for. Like every true hunter, he was not in favor of killing anything unless it was necessary. He asked the Lead, "Isn't there any way we can drive them off?"

"Not with the nests right there, Jason. It might be nightfall before we can get the job done, only to have the crocs reassemble under cover of darkness. You also need to realize that the eggs look like they're about ready to hatch. If they do, we could be facing an explosion of crocs."

"This has to be a group decision. We'll need to talk with the other safari leads, and see what suggestions they might have," Jason said. "We need to make sure that we make the right choice."

A Dead End

Alex listened as Jason and his team evaluated their options. He noticed that one of the younger hunters, Bill, was looking in the general direction of the crocs, but then shifted his focus to the nests and smiled. Finally, Alex said, "I have a feeling Bill might have come up with something. Let's think about it. Why is the croc herd there?"

Bill answered, "Because of the nests."

"Right you are," Alex replied. "How do you eliminate them?"

The Senior Lead jumped in with a reply. "First, we have to get past the herd that's blocking our way. But how are we going to do that?"

"Good question," Alex said, looking at the senior hunter. "What do you think?"

"I think we use the jeep with the shovel nose on the front," Bill said. "We can blow the horns to scare them away, and throw meat in the water to distract them at the same time. We should be able to get to the nests pretty easily that way."

Jason was pleased. His team had come up with a solution that didn't involve destroying the beasts. "Let's test it out."

Bill returned a little later. He was laying on the horn, and another junior hunter was ready with handfuls of raw meat. Within another half hour or so, the crocs had retreated to the water, leaving a clear path to the nests. After that the team made quick work of moving the nests, and returned to the main group, exuberant at their success.

"Your people did some great work!" Alex exclaimed. "Very creative thinking!"

"I know," Jason answered with a smile. "When we get to the clearing, we'll celebrate. They deserve some credit."

It was just after nightfall when the expedition reached the clearing. Jason was as good as his word, and once camp was set up, the party began. Under different circumstances, it might have continued until daybreak, but despite their jubilation at a job well done, the team members all knew that tomorrow would bring fresh toil and challenges, so they ended their celebration shortly after midnight. There would be no crusty eyes or headaches the following morning.

Early the next morning, a scratchy, static-filled call came through on the radio from the special detail. They were a half day's march from the clearing, and were presently coming up the same road that the rest of the expedition had followed.

"This is bad news," Jason said without emotion. "The shortcut obviously didn't work out, but at least they're safe."

Alex resisted the urge to remind Jason of his earlier warning, opting instead to reassure him. He knew all too well how useless – and even damaging – I *told you so*'s could be. "It was a calculated risk, and one that you thought was worth the investment."

"And I'd do it again under the same circumstances."

"Really?" Alex asked. "You'd do it, even knowing that you probably would fail?"

"Yes, because without taking some risks, I'd never learn what might work better than the things we already know how to do. And I'd always be tormenting myself with 'what-ifs.' It's that whole road-not-taken idea; that sort of thing drives me crazy."

"I understand that kind of thinking, Jason," Alex replied. "It's part of what has made your company so successful. I just hope that the next time, you'll have more data before taking that kind of chance. As it turns out, the maps you used to set your special detail's course were inconclusive. Better reconnaissance would have led to a better and less dicey decision. And that, my friend, is the difference between taking a risk and an outright gamble."

Jason's face had assumed a serious expression. "Don't get me wrong," Alex added. "You and your entire team have made significant improvements. You've already decided to make more changes before the next hunt."

Jason nodded.

"The question," Alex said pointedly, "will be how will you approach making those changes?"

Jason was puzzled. "What do you mean? I'll use the same approach we used for this hunt. The difference will be that we'll have a new set of problems to solve. But the approach should work for those as well, shouldn't it?."

"Well, let's think about that. We took out the so-called 'low hanging fruit' for this hunt," Alex replied.

"I'm not sure I follow you," Jason said, more puzzled than ever.

Stomp Yer Croc!

"It was your elimination of the obvious problems that made the difference." Alex replied.

"Hey, my *team* eliminated the problems. I didn't do it alone."

"Agreed, and they'll help you tackle the next set of problems that will be even harder to resolve. But here's the catch. Not all the problems you'll encounter will be so obvious. You may not even realize they are problems. Some young hunter may see issues that you, your leads, or senior hunters accept as normal, in spite of the possible consequences."

"So, you're saying that the next time around, we need to involve more junior people in solving problems," Jason said.

Alex nodded in agreement. "We can talk about this later. Right now, the highest priority is reintegrating the special detail with the main column."

It was nearly lunch time before the special detail reached the clearing. Alex, Jason, and the entire leadership team gathered together for lunch and to hear what happened.

"We'd seen the monkeys the last time I talked with you, Jason. They were up in the trees, and our team was paying no attention to them," the special detail's Safari Lead began.

"What changed?" Jason asked.

"Well, for nearly two days things went well. It was a clear road and an easy march. Then, things started to get bad. Monkeys and crocs were everywhere. Around that time, we realized that we'd lost communication with the main group. The monkeys got very bold. They climbed onto the team members' shoulders and stole supplies right out of our knapsacks. Some in the team tried making friends with one monkey, hoping he'd keep the others away. That didn't work, and only slowed our progress."

"I know from personal experience that monkeys can be a horrible distraction," Noel offered.

Alex and Jason looked at each other, nodded, and choked back laughter. Alex said, "Then what happened?"

The Special Detail Lead just shook his head. "You know where the telemetry picture stopped? Well, just as we reached that point, the crocs seemed to come out of nowhere. These were big fat ol' boys, too. They were snarling and snapping and as mean as they come."

Jason was astonished. "Was there any way around them?"

"No, sir," the Special Detail Lead replied. "There was a huge stone wall built across the trail, and the trail ended behind it. My people were shooting frantically, just to survive. As fast as the crocs submerged, more came from nests that were behind the wall. We were completely blind-sided."

"So, that's when you turned back?" Alex asked.

"Yeah. There's one other thing, though. The monkeys were running back and forth on top of the wall, obviously disturbed by the commotion. We were shooting at them with rubber bullets – didn't want to hurt 'em, just scare 'em – but as the bullets knocked them off the wall, the crocs ate them and – I swear – those crocs seemed to get bigger and more aggressive right before our eyes. It was a real feeding frenzy. It solved our monkey problem real quick, but the crocs became a bigger problem than they had been before."

"Hmmm... monkeys as croc food. What do you think of that, Noel?" the Senior Safari Lead asked with a grin.

Noel bristled at the teasing; it was clear that he didn't see the humor in it. Jason, Alex, and a few others found it impossible to suppress a chuckle, however. Then Jason looked at the expression on Noel's face and felt another small twinge of conscience. Noel really was trying to adapt, and he had made more progress than either Jason or Alex had expected

"Okay, let's lay off of Noel for now," Jason said, and he saw the glimmer of approval in Alex's eyes. "We have a few other problems to consider as well."

For Jason, the real problem was lost time. He was ahead of schedule, but knew that he'd lost at least another day waiting for the return of the special detail. Then, too, he'd sacrificed resources on that failed effort. Jason was beginning to suspect that perhaps Alex had been right about the search for the alternate route being a time-consuming, resource-wasting dead end.

Insights On The Team

After the special detail debriefing, Jason, Alex, and the Senior Safari Lead stayed to discuss the pros and cons of the failed effort. They analyzed the impact upon the overall mission, and reviewed the additional resources consumed and the time lost. "I've changed my mind," Jason said. "It wasn't worth the cost, after all. Next time, I'll be more careful."

"Never avoid a risk that has a reasonable chance of a payoff," Alex said. "The risks just need to be outweighed by the benefits and, then, managed and controlled. It looked like the special detail would provide a real benefit to the expedition, right up to the point where they ran out of options."

"I've been thinking about the valley," Jason said to Alex. "Cris Croc or no Cris Croc, that valley does have some pretty big crocs. I also think that when I was there before, I saw a wall like what the Special Detail Lead described. I'm concerned that we won't do any better than they did."

"I'll grant you that," Alex agreed. "But don't let it eat at you. You are learning and looking for improvement opportunities, but remember, everyone else is learning, too. Making – and learning from – mistakes is part of the hunt. As long as all of you share, compare, and leverage what you've learned, the whole safari will benefit."

"We still need to improve communications," Jason observed. "I'd have turned back before I got to the wall. There just wasn't any way to tell them to stop and return."

The Senior Safari Lead, who had been quietly listening, spoke up. "Jason, I've hunted with you for many years, and we've been successful. There's no question about it. Overall, the team has had good people on it. My concern over the years has been that sometimes we seem to have good people doing the wrong jobs."

Jason looked puzzled. "Over the years? Why didn't you tell me this before?"

Stomp Yer Croc!

The Senior Lead hesitated a moment, and then replied, "Well, since we're being frank here… in the past, you weren't so receptive to others' ideas, and I've been hesitant to let you know my thoughts. I didn't want to overstep my bounds. But this hunt has been different. It seems like you're looking at us as a real team now, instead of just employees, hired to perform specific tasks."

Jason thought a moment, and realized that the Senior Lead was right. "I understand. It's always been my way or the highway, and you all deserved better. Heck… even I deserved better. I wonder how much more successful our previous hunts could have been if I hadn't been so… well… stubborn. But getting back to your point about having people assigned to the wrong jobs, what do you mean?" Jason asked.

"Most of Team 2 is made up of grasslands hunters. None of them has the skill or knowledge for a jungle hunt, much less a swamp. They're basically rookies, when we've really needed seasoned team members. True, they've been learning over both of the hunts, but that education has cost us, probably more than we think."

"So, you think we should have hired new folks?"

"In hindsight, sir, yes, I do," the Senior Safari Lead answered. "Of course, now they're catching on. Next swamp hunt, they'll be more productive, but getting up to speed takes time. You paid the price of their learning curve, and it cut into our success."

"He has a point," Alex commented. "When you're outlining your next 'Lessons Learned,' the team might look at better ways of choosing or training staff for future hunts. The kind of hunt and the territory to be crossed are just two of the things you'll need to think about."

"There are some major differences between grassland and swamp hunts," the Senior Safari Lead continued. "Some things, like using and caring for a machete, can be taught. Those are easy skills to learn. How to deal with snakes and crocs in the water is another matter."

"I think we still need to groom young hunters," Jason objected.

The Senior Safari Lead responded, "True enough for the young hunters. Their jobs aren't as crucial to the overall success of an expedition, and the basic skills they need can be taught. But our team leads and Safari Leads need to have the right kind of experience going in, rather than being expected to acquire it along the way."

"I agree," said Jason, and Alex nodded in agreement, as well.

"Here's something I want you to do when we return," Jason said to the Senior Safari Lead. "Set up a chart by type of hunt, listing individuals' responsibilities and skill requirements. It will take some time, but in the long run, it will help us to better plan future hunts."

An idea crept slowly into Jason's mind as he thought about his conversations with Alex and the experiences of this hunt. "I think I understand about 'hot spots,' Alex," he declared. "We're dealing with problems one at a time."

"And?" Alex asked.

"Some problems are related to others. If we group problems, we can solve larger ones!"

"Good point… but actually, they all are related," the Senior Safari Lead observed.

"I know," Jason replied. "But we need to sort them into groups we can work with. After this hunt, we're going to try it the new way. No more onesies, twosies on our problems."

Jason pulled out his notepad and started to write.

"Want some help?" Alex asked.

Jason smiled. "You bet! The more hands the better."

Later that afternoon, Jason and the Senior Safari Lead presented the idea to the other Safari Leads and team leads. "Start making a list of problems you encountered, including the ones you've given me already. Talk with your junior hunters, too. They're going to get involved."

The leadership team was a bit surprised by the involvement of the junior hunters. Jason explained that involving junior members should prove useful in the "Lessons Learned" segment of the debrief, as well as in other activities.

"When we finish this hunt, we're going to group these problems and figure out which ones are most closely related to each other," the Senior Safari Lead continued. "But be ready for a 'Lessons Learned' session and a review of our tracking numbers first. It all works together."

* * * * *

Later, as he sat alone, with thoughts of the hunt rolling through his mind, Jason felt a new sense of calm, and a deeper confidence than he had ever known. It was much more than feeling assured of his own abilities, which he had never doubted (at least, not out loud). Rather, it was a wholly new level of awareness. The success of his hunts – and of his company itself – would be bolstered by the fact that his own abilities would be compounded by those of everyone who worked with him. Even more reassuring was the newfound level of respect and trust he had for those people whom he had so long seen as extensions of his own ideas. They had ideas of their own, and good ones, at that.

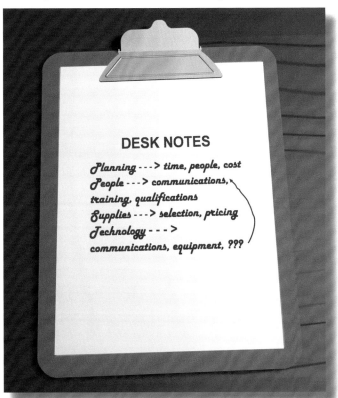

Now, perhaps for the first time, they would be made to feel that their ideas were crucial to the hunts in which they participated. And they wouldn't just be participating; they would be partners in the efforts, and real owners of every success they achieved.

It was a good feeling that Jason knew all too well, and that he found himself wanting – no, *needing* – to share.

four

Gauging Progress

Jason had originally planned to stay in the clearing for only two days. With the delayed arrival of the special detail, however, the rest period was extended to four days. He didn't like losing the time, but he was well ahead of schedule, and the special detail needed to rest. They also needed to be re-integrated with the main team after their ordeal.

Furthermore, Jason felt that they needed to be given the main team's newfound knowledge and experience in dealing with the crocs. Since there was so much information to be gathered and evaluated against the plan before the re-planning meeting could begin, the group decided to delay the meeting by one day.

Jason was keenly aware of how differently they were doing things under Alex's tutelage. As much as he didn't like to admit it, on previous safaris the written plans had often ended up in the glove boxes of the Jeeps, gathering dust for the entire journey. Any analysis of the team's actual versus planned activities had been very informal. It mostly had been done in the leadership team's heads, if at all. *Those days are gone*, Jason thought to himself, *and good riddance*.

Most of the leadership team believed that the expedition was well under budget and comfortably ahead of schedule. The biggest question was the impact of the special detail on the overall progress of the expedition.

As it turned out, however, the delay caused by waiting for the special detail had been used to great advantage, as it had offered the team leaders in the main group an opportunity to review the original plan with their teams, and to measure the expedition's actual progress against it.

The individual teams had reported their findings to their Safari Leads, who were then able to assemble a pretty accurate overview of the entire effort.

The Senior Safari Lead had then coordinated the efforts of the other Safari Leads and their Team Leads. Although he hadn't rested much himself, he'd made sure that everyone else got as much rest as possible. As a result, everyone had proceeded at a much more relaxed pace than was possible when they were on the trail. Hence, there was an upside after all to the delay caused by the special detail.

During their "down time," Alex had focused on working with the Safari Leads and their senior hunters in small groups, determined not to upset the organizational balance that Jason had created. If he had a concern or a recommendation, he gave it to Jason privately.

For his part, Jason had scouted the trail ahead, already anxious for the next foray into the valley. Getting into the valley, rather than merely lingering at its entrance, remained his ultimate measure of success.

Measuring Progress

When all the data were finally collected, the numbers looked good – not as good as everyone had hoped, but still a major improvement when compared to the first hunt.

Although Jason had participated in specifying the kinds of measures he needed for decision-making, the format for the final report had been designed by his senior staff before the second hunt began. It included several charts, the first of which compared remaining resources for the first and second hunts at the time the safari had reached the clearing.

The fuel supply was a big concern, as the team intended to recover the personnel carrier and other vehicles that had been left behind on the first hunt. The Senior Safari Lead warned, "Recovering the personnel carrier and these other Jeeps might force us to leave a few Jeeps here, just to ensure enough fuel to get out of the swamp."

"Team 3 isn't really needed in the valley," Jason said. "If they stay behind, we'd save enough fuel to let us salvage those vehicles and recover some of the lost cost from the first hunt."

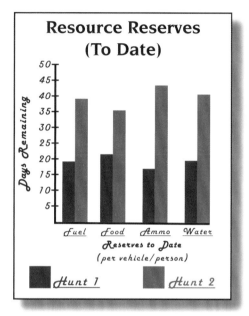

A senior hunter from Team 3 said, "Those Jeeps will need some repairs before they are fit to use. While you go on to the valley, we can work on them."

"Okay, it's agreed," Jason replied. "We have enough Jeeps and people to get the gold out. With the personnel carrier recovered and repaired, we'll shift the load to it when we return."

The Senior Safari Lead continued, "We've hardly used any ammunition compared to the last hunt. Mainly, it's because there were fewer crocs on this hunt, and those we've seen have been much less of a threat than we'd expected. Plus, we've found other ways to handle problem crocs."

"Well, we used a lot of ammunition shooting at monkeys at the wall," the Special Detail Lead interjected. "For what little good it did, we probably accounted for most of the ammunition that's been used."

"True enough," the Senior Safari Lead granted. "In spite of that, we have three times more ammunition at this point than we did on the first hunt. We may need it in the valley."

"Let's hope not," Jason said. "With any luck, there will be fewer crocs in the valley than we saw before."

"We can't count on that," Alex reminded him.

"Food was the big issue on the last hunt. I'm determined that no one will need to go on reduced rations like last time," the Senior Safari Lead said. "We have twelve days to get to the valley and return to this clearing. Then, we'll have a little more than four weeks' food supply to get out of the swamp."

Alex spoke up again. "That assumes the current level of food and fuel use. In any case, leaving Team 3 here with a fuel reserve makes sense. If the reserve is still here, then you can get the old Jeeps and the personnel carrier out. If not, at least all of your other Jeeps are safe."

"What is the level of fuel use?" Jason asked.

A senior hunter replied, "We made improvements to engine tuning and idle speeds for the second hunt, and those have helped a lot. With the Jeeps moving at a better-than-average speed, they're not sucking fuel like the last trip. We're also doing more miles in a day, so we've been on the trail for fewer total days."

"The result is better overall fuel mileage," another of the Safari Leads added.

"It doesn't hurt that those on foot are moving more quickly now, either. That enabled the Jeeps to move at a faster pace along with them, and the Jeeps were better able to help those on foot by carrying those who got tired or were moving more slowly than the main group. A good trade-off, I'd say," Alex offered.

"Thank you," Jason said, looking around at the group with a big smile on his face. "I'm pleased with and proud of each of you. It was your improvement efforts that made this happen. Let's meet tomorrow and settle on any other changes we need to make."

Mid-Course Corrections

Jason sponsored another smaller celebration after the progress meeting to reward the entire team for their good performance. The hunters were all in good spirits, and looking forward to the final leg of their journey.

"I still have questions about the changes in fuel use. Any chance that the conditions will change?" Alex asked Jason.

"From here to the valley, it's better terrain," Jason replied. "Fuel usage rates should get better, not worse."

"Does that mean the men on foot will move faster, too?"

"Actually, it does. On the last trip, they made about two miles more per day, and the Jeeps performed accordingly," Jason said. "No reason to believe that those conditions will change. Besides, I've scouted ahead, just to be sure."

Alex found no reason to challenge Jason's assertions, knowing that his decision was based on experience and good, sound numbers. In spite of this, Alex knew that Jason needed to expect the unexpected.

As planned, the meeting to discuss additional changes didn't begin until later in the morning. Since the party had lasted well into the night, Jason knew that everyone, himself included, needed some extra sleep. When the team finally assembled, the members did indeed look refreshed, if a bit bleary-eyed.

The meeting began with a suggestion from a spokesman for the Safari Leads. "We'd like to leave two Jeeps at the end of the column and one in the lead. All of the remaining Jeeps will head straight for the valley at full speed – an advance party, if you will."

"Why?" Jason asked.

"Several reasons, sir," he said. "First, driving without the stop and go of moving with the column will increase gas mileage. Second, this advance party can establish a base camp to save time later. Finally, they can evaluate conditions and begin taking any necessary measures. This, too, may save time."

"That sounds like a great idea to me, and I can see no additional risks," Jason replied. "If all of you are in agreement, then go ahead. I want the Senior Safari Lead to go with you. He can act on my authority. Make certain you don't leave until we've cleared the swamp. By delaying your group's departure, the rest of the team is only a day-and-a-half behind you."

"Thank you," the Senior Safari Lead said.

"I'd like to add a recommendation," Alex said.

"Have at it," Jason replied.

Alex stood up. "If you see a problem, and you can't raise us with the trail phones, send a Jeep back. If it's too serious, all of you need to return to the main column."

"Good point," Jason added. "We don't want another experience like the special detail had. No sense tackling any big problems without the rest of the team."

The Senior Safari Lead agreed, as did the spokesman. They all understood that the expedition did not want another risk to become an issue.

Jason continued the meeting with a question, "Are there any other suggestions?"

"Do you still plan on going deeper into the valley this time?" Noel asked.

"Yes," Jason answered.

"Then, will the base camp be at the entrance to the valley or inside the valley?"

"I'd say that at the entrance would be safer," Alex responded. "We don't know how many crocs will be there."

When a younger hunter raised his hand to speak, Jason nodded in his direction. "Isn't it odd that, on our last expedition, while there were no crocs between the swamp and the valley, there were so many crocs actually *in* the valley?"

Jason, the Senior Safari Lead, and Alex exchanged questioning glances. The question, at first, seemed naïve. Noel started to speak, thought better of it, and stopped. None of them had really given the matter any consideration. It seemed too obvious.

Finally, the Senior Safari Lead broke the silence. "Leading up to the valley, you have mainly hills and rocky ground. Granted, there are some small marshes and grassy areas, but it's no place for crocs. There are, of course, lots of snakes and some harmless little critters."

"The valley, however, has marshes and rivers where the crocs can thrive," the Senior Safari Lead continued. "I can't think of any reason for the crocs not to be there."

"What I'm trying to say, sir," the younger hunter replied, "is what if the crocs get behind the advance group? They'd be trapped, and we wouldn't be able to get through without their equipment. It's not as if we have an alternate route. The ground between the swamp and the valley is where the mountains and cliffs begin. Once we're out of the swamp, it's the only path we can take."

"So, you think it's safer to stay together and move more slowly," Jason said.

"Yes, sir."

"Once we're out of the swamp, it's the only path we can take."

Alex uncrossed his ankles, shifted his feet, and stood up again. "He has a point. The crocs can be unpredictable. You know that they can be there waiting for you where you'd least expect them."

Stomp Yer Croc!

"The Safari Leads seem to think that the advance group is a good idea, and the risk of the crocs coming out of the valley is low," Jason said. "Once we're actually in the valley, though, it's a different story."

Jason had not directly addressed the young hunter's concern, and the latter sat down and didn't pursue it any further, though he remained worried. The Senior Safari Lead could see that the younger hunter was still concerned and decided he'd talk with the man after the meeting.

Jason had hoped for more discussion and suggestions for improvements, but the team's focus was on finishing the hunt. The suggestions they did provide were related to getting things moving quickly, and they pretty well overlooked the last phases of the hunt. Sensing the team needed more rest, Jason decided to cut the meeting short and re-convene the next day to finish up the "re-planning." They could discuss those future improvements at that time.

An Easy Re-Plan

Unlike the previous day's meeting, the second day's re-planning meeting began early. "We've looked at our success measures and our progress to date. At this point, it looks like we can afford to spend one extra day in the valley," Jason began.

"That assumes that we hold to the schedule to reach the valley," the Senior Safari Lead added. "We adjusted the order of the Jeeps so that, as soon as we are out of the swamp, they can move ahead and get things ready for us."

Another Safari Lead said, "The new schedule is ready, and each of the team leads has copied it down. If anyone has any questions, have them talk with their senior hunter."

"Make sure that everyone has a sharp machete!" Noel shouted. "The last bit of swamp is pretty dense."

Jason's leadership team had made very few changes in the plans for the final leg of the hunt. Most of the changes allotted additional time to the different activities in the valley. The schedule for actually getting to the valley was tightened up by using the data they had on progress to date.

"We spent a lot more time re-planning here than we did on the last hunt," Jason remarked to Alex.

Alex nodded. "Not only did you spend more time preparing on this trip, you spent more time resting – and still you lost less time."

"We stuck to the plan this time."

Alex smiled. "I call that 'walking the walk.' You did what you said you'd do."

"On the last hunt," Jason mused, "everyone had a copy of the plan with them, but no one used it. Now, each of the senior people checks the plan and schedule every morning."

Again Alex nodded, adding, "I've been thinking about something else, too. That advance group you're going to be sending… you know, that's a very good idea for another reason that no one mentioned."

"What's that?" Jason asked.

"Going through the swamp successfully is only the journey to the valley. It's *in* the valley that you'll find out if the trip was justified."

"Well, yeah… but only if I find enough gold," Jason replied. He was puzzled about why Alex would take such pains to state the obvious.

"It's more than just finding gold," Alex said. "It is only after you get into the valley that the people who want what you can deliver will be satisfied. The banks and financiers will know if it's worth it to keep supporting you on future hunts. Lastly, the team will have the satisfaction of being part of a highly successful hunt."

"Okay, I get the deeper lesson here – and point well taken – but from a practical standpoint, what does that have to do with the advance group? They won't be bringing out any gold. They need to wait for the rest of us."

"They'll do more than wait. Their job is to find ways of improving on the team's efforts and enhancing the level of success," Alex responded. "It's not just the quantity of gold that determines your success. The quality of the gold establishes how much it – and your journey – are worth."

With that Alex said good night and went to his own tent, leaving Jason with some new ideas to mull over. Jason had a feeling that once this trip was completed, neither he nor his team members would ever look upon the process of hunting for gold in quite the same way.

Following Through

The next morning, the expedition left the clearing, feeling prepared and confident. A small team used their freshly sharpened machetes to cut through the remaining undergrowth as the advance team's Jeeps edged along behind them. They were making steady progress at the expected pace. There wasn't a croc or a monkey in sight, and the entire team was able to focus upon leaving the swamp and getting onto the main road to the valley. After what felt like an eternity (especially to those charged with the task of chopping their way through the underbrush), the swamp finally gave way to the more open marshland. A few hundred feet ahead, the beginning of the gorge and the road leading into the valley were visible.

"We're finally out of the swamp!" Jason shouted. A hearty cheer arose from the team.

With the machete carriers standing at the side of the road taking a much-deserved rest break, the advance group's Jeeps sped off into the gorge. As the pace of the main column also increased, Jason found himself able to relax a bit. He expected their progress over the next two days to be steady and relatively easy.

Meanwhile, the Senior Safari Lead, riding in the Jeep at the head of the advance group, took off down the hard-packed road as quickly as the driver could safely go. At some points, they had to slow down to avoid rocks that had fallen onto the road. After one turn in the road, they came upon a clutch of snakes basking in the sun. Feeling the vibrations of the Jeeps on the road, the snakes quickly slithered away into the patches of grass amid the rocks.

The Senior Safari Lead didn't like slowing down, but felt he had no choice but to do so. *We don't need a wreck now*, he thought to himself. *I've got the responsibility this time, and I don't have the luxury of needless mistakes.*

All of the drivers left plenty of room between their Jeeps and those in front of them, a rule upon which Jason had insisted. As a result, the advance group was spread out such that the Jeep at the back of the line often couldn't see more than two or three Jeeps in front, especially when they were passing along a particularly dusty stretch. Those near the end of the column often were choked by the swirling, sandy powder, and some actually began to feel nostalgic for the muddy paths through the swamp.

Just before they reached the entrance to the valley, the Senior Safari Lead halted the advance group. Slowly, one after the other, the Jeeps caught up and came to a stop. "I want everyone to take a breather before we get to the valley," the Senior Safari Lead said. "Let's review the plan while we eat lunch. I want to be sure everyone knows what to do."

Looking at everyone's dust-covered faces, Noel added, "You chaps might want to clean up a bit, too."

There were few questions as the Senior Safari Lead reviewed the plan. Most of the concerns dealt with timing rather than responsibility, since all of the team members knew their roles.

After a one-hour break, they climbed back into their Jeeps and took off toward the valley entrance. An hour later, they arrived and halted, awaiting further orders. The Senior Safari Lead and two senior hunters drove a little farther to take a peek at what lay ahead.

They found a good spot to park the Jeep, and got out and stretched their legs. The Senior Safari Lead raised his binoculars to his eyes, scanning the valley for a moment before stopping short, his jaw dropping at what he saw. "Well this is different!" he exclaimed as he stared through his binoculars. "If Jason doesn't believe in old King 'Cris' Croc after seeing this, he never will. I'm calling him ASAP!"

No Stroll To The Finish

While the advance group moved quickly, Jason, Alex, and the main column proceeded slowly and steadily. The Senior Safari Lead had called him when they stopped for lunch. So far, everyone had followed the plan.

Jason had also stopped to rest his team, but had lingered for only thirty minutes, compared to the hour the advance column had taken. With his people on foot in the heat, he let them rest after each grueling 45-minute march, and his leads and senior hunters kept an eye on their fresh water reserves. "No reason to die of thirst," Jason remarked.

"We have plenty of water," a Safari Lead answered. "The team is following the guidance to the letter. A half-cup at each break. So far, no one seems dehydrated."

"That's good, but watch the water reserves just the same. We've done well to this point. All of us agreed to be ready for an emergency. Let's stay with what we said we'd do."

The main column marched another hour and was just starting again after a rest when Jason heard from the Senior Safari Lead again. "Jason," he said, "The valley looks different."

"Different? How?"

"Well, that depends. There seems to be a lot more gold than what we saw before, and, judging by the color, it looks to be more pure," the Senior Safari Lead began. "And this is really weird… but the crocs are a lot smaller… except for one."

"Are you starting that King Croc nonsense again?" Jason asked, somewhat sarcastically.

"Yes, sir, I am, but I assure you that it's not nonsense. You'll just have to see it for yourself."

"I will, and I'll put an end to this myth fast. What else do you have to report?"

"Jason, I wasn't sure about this because of the low clouds at the end of the valley. So, I went back and got the Lead for the special detail. He's not certain, either, but there may be another big stone wall way back there near the cliffs."

"Is there anything behind the wall?"

"Can't tell because of the clouds. When the main column comes up, we can try and take a closer look," the Senior Safari Lead answered.

"What's your team doing now?"

Stomp Yer Croc!

"Building a gated *chevaux-de-frise* and putting the Jeeps on our side of it. However, if those crocs come at us before we're done, we'll have a terrible time of it."

"So, there's no immediate threat?" Jason asked.

Somewhat surprised, the Senior Safari Lead answered, "It depends on what you mean by an immediate threat!"

Jason was losing his patience, which he knew was something he couldn't do. He took a deep breath before asking, "Are you sure you're not overreacting? Is a *chevaux-de-frise* really necessary at this point? There really shouldn't be any reason to panic."

"This is the safest way to go, Jason. And no, I'm not overreacting. We need to be prepared." The Senior Safari Lead hesitated before continuing. "I think we have the strong possibility of a threat. These crocs are mostly smaller, but there are hundreds of them."

"All right, then. Stay in communication. We'll be there tomorrow afternoon as planned." With that, Jason signed off rather abruptly. He was anxious to catch up to the other group and quell this King Croc nonsense before it spread to the entire expedition.

Surprises In The *Valley*

When the main column reached the advance team's camp, Jason wanted to go immediately to the valley to see what the Senior Safari Lead had been so concerned about. Alex and Jason joined the Senior Safari Lead and the rest of the leadership team, and a few of the senior hunters followed from a distance.

When they reached a vantage point, the Senior Safari Lead handed the binoculars to Jason. "Look over there first. See how many more crocs are in the valley this time? Now take a look at the big boy near the end of the valley, by the gate in the middle of the wall."

Jason scanned the middle of the valley floor. "My word! Look at the size of that monster! That's the biggest croc I've ever seen!"

The Senior Safari Lead said nothing; "I told you so" wasn't in his repertoire, and he wouldn't have said such a thing to Jason anyway. He watched Jason's expression as Jason looked toward the far end of the valley, staring through the eyepiece and straining to get a clear view.

Finally Jason spoke in wonderment, "Those aren't clouds creating too much of a haze; it's the glow of the gold that's obscuring our view of the wall behind that big croc."

"My word! Look at the size of that monster! That's the biggest croc I've ever seen!"

"The big question is, how do we get close enough to really see what's there?" the Senior Safari Lead said.

"We'll meet and figure it out," Jason replied. "After what we've gone through just to get here, I'm not leaving anything off the table."

As Jason and Alex walked back to camp alone, Jason seemed contrite. "If that big croc is King Croc, then I've obviously been blind for a long time."

"What's the old chestnut? Seeing is believing?" Alex said.

Jason chuckled. "Sure is. You could throw out a few more about old fools and teaching old dogs, but spare me. I already know that you and a lot of other folks were right."

Alex said, "This is the most serious challenge of the hunt. A bigger one even than the special detail. So let me offer one more cliché."

"Go ahead. What is it?" Jason asked.

"Don't fall on your sword."

"Don't plan to," Jason said, more flippantly than he'd intended.

"I'm serious, Jason. You can't beat King Croc on this hunt. The plain truth is that the rifles you brought simply aren't powerful enough to kill him, and he doesn't look too inclined to run away in fear. The best you can do," Alex advised, "is to get what you can without actually confronting the old boy. I wouldn't feel too badly about it, though. There's plenty of gold to be had, and lots more than on the last trip, from what I can see."

"If my Senior Safari Lead is to be believed," Jason replied, "there's not only more gold, it's better quality, too."

"Your bankers and financiers will be more pleased than they were the last time. So will the folks who'll buy what you bring back."

Jason nodded in agreement. "Let's get back to camp. We need to re-plan our approach to the valley."

Jason was excited, perhaps more excited than he had ever been on a hunt, but his excitement was tempered by the knowledge that the stakes were higher than they had been on previous hunts. The quest for gold, once so seemingly straightforward, had reached a whole new level of complexity. Jason, consummate hunter that he was, still found himself wondering whether he – and his team – were up to the task. One way or the other, he was about to find out.

Stomped Crocs!

Early the next morning the entire team gathered at the entrance to the valley, which was now visibly swarming with crocs; there was no longer any debate about that, anyway. Gazing at the teeming mass of reptiles, a Senior Hunter offered, "We have plenty of ammunition. Why not just go in with our guns blazing and shoot any of the crocs that come close? If need be, we can kill the lot of them."

"For one thing, it isn't humane unless we're left with no alternative," Jason said. "In addition, they'll just hatch more, and we'd just have to deal with them on the next hunt."

"We've shot them before," a Safari Lead argued.

Jason responded slowly and carefully. "Actually, we've shot *at* them. It does no good. They just go away and re-appear somewhere else. We need a better solution than brute force."

"Let's think about what's worked so far," Alex interjected. "We moved the nests, and that worked. Another time, we diverted their attention while we eliminated the nest, and that worked."

The Senior Safari Lead asked, "What kinds of distractions work the best?"

"Mostly depends on the croc, I'd say," Noel replied. "Each croc, or maybe, each croc herd, seems to act differently."

Alex agreed. "That's a sound generalization, Noel, but we need to find a specific solution for these crocs. Our other experiences were with swamp crocs. I know that this is a generalization, too, but these are valley crocs, and from what I've seen so far, they seem to behave differently."

The Special Detail Lead offered an observation. "Back at the wall on the dead end road, the crocs seemed to be primarily interested in stopping us. Yesterday, I noticed that these valley crocs acted like they were more interested in holding on to the gold. I'm not sure, but that may be one clue."

"What about the nests?" Jason asked.

The Senior Safari Lead answered, "They are farther away than those in the swamp were. One thing that seems to be the same is that the more attention we give to the crocs, the more rapidly the eggs hatch, until the nest is empty."

Alex asked, "Then what happens?"

"In the swamp – and the last time here in the valley, too – the crocs went away after that initial swell in their population. Moving the nest in the swamp the first time was a mistake, since it only postponed dealing with the problem." The Senior Safari Lead paused for a moment. "You see, once all the eggs hatch and the nest empties, we have a known quantity to deal with, and the crocs can get stomped."

"Then the key in the valley is 'attacking' the larger crocs and charging the nests to stomp the eggs and any baby crocs. Right?" Jason responded. Several team members nodded.

"That still leaves us with the question of how to attack the crocs," the senior hunter said. "Shall we just go in and stomp them?"

"That sounds easy when you say it real fast," Alex said, smiling. "But it's not that simple. Remember, we only need to get the crocs out of the way long enough to eliminate their nests. That's the only way to really get rid of them. Knowing what kind of crocs are being hatched, we can adjust our attack accordingly. No nests, no more crocs."

Some Crocs Retreat

Jason had the Safari Leads organize small teams to test out some of the different methods of dealing with the crocs. The results were mixed. Some solutions worked with some crocs and their nests, but not with others. Other solutions only seemed to work with a few croc herds, and not at all with the others. Of course, these were only tests, and naturally, Jason and Alex were most concerned about the first group of crocs with which they would be dealing. It was important to find the best solution, and the sooner, the better. At last, the tests had been completed, and with the reports from the teams in hand, the Safari's leadership met again to finalize its initial foray into the valley.

The Senior Safari Lead expressed concern that their tests had only dealt with the crocs that were in the middle of the valley. "What about King Croc and that herd near the cliffs in the back?" he asked.

"Let's take on what we can manage first," Jason said. "If we can get into the middle of the valley, we'll move the base camp forward and hold the ground we've gained. I don't want to fight the same group again the next day, or even on the next hunt."

"What do we do once we've reached the middle of the valley?" the Senior Safari Lead asked.

Jason had anticipated that question. "The same as we did this time. We'll test our solutions and stomp the next wave of crocs and the next, until King Croc and his boys are the only ones left."

"But we only have four days to gather the gold and begin our return trip."

The rest of the team listened quietly as Jason and his Senior Safari Lead continued their exchange.

"We'll take and hold as much ground as we can until those four days are done," Jason said. "On the next hunt, we'll be ready to stomp even more crocs."

"You won't give up, then, no matter what kind of crocs or whatever else we run into," said the Senior Safari Lead, and it was more a statement than a question. Actually the Senior Lead had worked with Jason long enough to know that he would never back down from a problem until it was completely resolved. Halfway measures just didn't exist in Jason's world. Some of the younger hunters had been looking a little doubtful during this exchange, and the Senior Lead correctly sensed that maybe this was a good time to sneak in a motivational lesson or two – kind of a mini-pep talk.

Jason took the cue like a pro, smiling his approval at his Senior Lead man. "You got that right," he replied. "Giving up is not an option. If we persevere, every hunt will get easier, and the results will get better. What I've learned, and what all of us should have learned by now, is that solving a problem once isn't enough. We need to make certain that each response to a difficult situation continues to work on each and every hunt. That way, we prevent problems from reoccurring."

Stomp Yer Croc!

"Otherwise," Alex broke in, "you'll repeat the bad experiences of the first hunt all over again."

This exchange between the two leaders helped iron out last-minute problems and questions, and had the secondary effect of motivating all of the Safari Leads and Senior Hunters for the journey into the valley. Jason felt good about what they'd accomplished at this "mini-meeting." Now it was time to put it all to the test.

As the expedition entered the valley, the group split up into teams that went after specific croc herds. Each team applied the pretested methods for the herd that it needed to stomp. Some croc herds just disappeared completely. To the surprise of some teams, however, some of the crocs didn't attack, nor did their nests completely empty out. Instead, more eggs materialized, and the crocs dragged their nests deeper into the Valley, to the security of King Croc's domain. In the distance, Jason watched as King Croc appeared to spontaneously grow in stature, and the amount of gold behind him redoubled.

Stomping Crocs!

By the end of the first day in the Valley, Jason's expedition had collected as much gold as they had during the entire first hunt. Jason felt satisfied with the first day's effort, and felt ready for whatever challenges the next two days might bring. Even though the entire team was feeling jubilant about the first day's success, Jason resisted the urge to hold a celebration for the team. That party would wait until their last night in the valley.

The following morning was spent reviewing what had happened the day before and refining their methods of dealing with the crocs. Then, another group of teams went out to test the refined techniques on the remaining herds. Once again, they met with mixed success; their methods worked about 80 percent of the time. All of the small teams reported that this time, there seemed to be one or two large crocs, surrounded by many smaller ones in each herd.

"Something odd happened with our team," a senior hunter said. "When we'd stomped all the smaller crocs, the big croc and the nests all disappeared."

Another senior hunter offered that his team had the same experience. "The eggs didn't even hatch! We were waist high in the marsh and suddenly it was dry land. No crocs and no nest to be seen anywhere. It was the easiest stomping we've done in two days."

"What about King Croc? Did he come near your team?" Jason asked.

"No, Jason, he didn't. His herd got a little larger and so did he. Yeah, I know that sounds strange, but I know what I saw. Anyway, old Cris Croc stayed put, right by the big stone wall."

By now nothing was surprising to Jason. Big crocs, little crocs, crocs that seemingly changed size at will… maybe the folks back home wouldn't believe it, but Jason had no trouble accepting his senior hunter's account, because he had witnessed the same things himself. If there was one thing he'd learned, it was that expecting the unexpected was the key to understanding this strange place. No matter what happened, he felt as if he and his team could handle it.

"Let's make the corrections we need," he said calmly. "Then, all of us can go in again this afternoon. Let's see how we do this time."

In general, the second day was easier than the first. Fewer crocs were encountered than before, and more nests were eliminated, along with the croc herds surrounding them. The marshes also receded.

As the Safari Leads and their teams battled the crocs and pushed deeper into the valley, they drew closer and closer to a ridge. Before long, they began to see King Croc on the other side of the ridge; he was not completely visible, but they heard him roaring and thrashing around in the marsh water. Finally, King Croc became the challenge that had to be faced.

The teams advanced more slowly now, both out of caution, and because they had begun to tire from the exhausting work. The hunt eventually came to a complete halt at the foot of the ridge. With dry land now behind them, the base camp was moved forward again.

As for their reason for making the dangerous trek in the first place – the gold – well, that was the good news. The amount of gold that they got was twice that of the day before. Jason now estimated that the final return would be almost five times greater than that of the first hunt.

Stomp Yer Croc!

Because the second hunt had cost less in labor, time, and resources, the profit margin had more than doubled. "This hunt is definitely a success!" Jason declared, secretly amused at his own knack for understatement.

*"**This hunt** definitely is a success!" Jason declared.*

It soon became clear, however, that the success realized thus far wasn't quite enough – not for Jason, anyway. Although Alex and the Safari Leads believed it was time to end the hunt and return to Supply City, Jason wanted to stay a while longer and face off against King Croc. Where once he had denied that this croc even existed, he was now determined to defeat it. In fact, much to the visible dismay of some of the team members, he had no intention of turning back until he'd at least had an opportunity to face and overcome this last challenge.

Is More Gold The Only Goal?

After Jason had announced his intentions, Alex pulled him aside. "Jason, you and your team really aren't ready yet," he said. "That's all right, though! After all, the goal was only to improve on the first hunt. It was to make things better and to improve on your limited success on the last hunt."

"It's not just about the gold, Alex. It's also the journey – and the challenge."

Shaking his head, Alex tried to explain what he meant. "Cris Croc is no ordinary challenge. He's a belly full of big problems all rolled into one oversized critter. The crocs in his herd are big, too. Wait until the next hunt."

"I'd like to try anyway, Alex. We need to experiment to see if this Cris Croc can actually be defeated."

So it was that on the fourth day, rather than heading back to Supply City, the teams reluctantly went over the ridge to test their solutions against King Croc. Less than two hours later, they returned exhausted and discouraged. "None of the techniques worked," said the Senior Safari Lead. "Every nest hatched all of its eggs, and new eggs appeared. We need new solutions. In fact... well, we agree with Alex." It was clear that the Senior Lead was a bit hesitant to admit the last part.

Jason was silent for a few moments, thinking. He finally asked, "Well, did you stomp any of the crocs?"

"A few, but the marsh stayed flooded, and new crocs took their place."

Stomp Yer Croc!

"Let's go to the crest of the ridge," Jason said to Alex and the Senior Safari Lead. When they arrived, they stood at the top of the ridge staring in disbelief.

"There's four times as much gold here as we've taken in the past two days!" Jason shouted. "Plus, there are more crocs here than we've already defeated. It will take more time and effort than we have." He struggled to keep his disappointment from showing.

Alex sensed it anyway, and turned toward Jason to reassure him. "You've done well, Jason, and so has everyone on your team. You just need to take some time to figure out how to make the other improvements that your team will need."

Jason thought about it a moment, and then nodded. "Yep, got that right. Thanks, Alex. Let's go back to camp. I need to speak with the team."

Even Jason now knew that it was time to return to Supply City.

The four men climbed into the Jeep and drove down the slope. By the time they got into camp, Jason's disappointment had abated, and he instructed his leadership team to set up for a celebration that night.

After dinner, Jason stood up and banged his spoon on a large tin plate. Everyone stopped talking and looked at Jason.

"This was a very successful hunt," he began. "You've improved and, hopefully, so have I. Let me apologize to those who made the sacrifice this afternoon, chasing after solutions to the King Croc problem. That was my mistake. You did your jobs as best you could against nearly impossible odds."

Much to his surprise, the team gave Jason a standing ovation. As they clapped their hands and cheered, Jason just stared in amazement, his eyes moistening quite against his will. He wanted to say more, so he raised one hand until the team sat down and grew quiet again.

"Thank you," he continued. "I'm not sure that I earned that kind of appreciation. I do have a surprise for all of you. When we return to our home base, I'm going to instruct our accountant to increase your scheduled bonuses by ten percent. You most definitely have earned it with your performance on this hunt."

Once again, the team stood and applauded. "Let's pack in the morning and head home," Jason concluded. "Thank you again."

It wasn't an eloquent speech or a long one, but Jason had said what he needed to say. Even Noel gave him an approving smile. H*ow much he's changed*, Jason thought. O*r maybe he's the same, and it is* I *who have changed*. Or *maybe both of us*. It was clear that Noel fit in better with the team now, in any case.

After the celebration ended, Jason returned to his tent. As the rest of the camp slept, he sat at his table writing furiously. He could not sleep; there were so many ideas that he wanted the team to explore.

It was well past midnight when he finally stood up and stretched. He still did not feel sleepy, however, and found himself walking out of the tent and toward the ridge. It wasn't long before he was standing again on the crest – alone, or so he thought. Staring through the mist, he could see golden reptilian eyes glowering at him from the mists before the wall. "What a magnificent challenge you will be," he said aloud.

"He's more than a mere challenge," a voice said from the darkness. "He's a *problem* – possibly a lot of problems you have yet to even imagine."

Jason was startled, but he recognized the voice, as well as the sentiment.

"Alex, what are you doing still up?"

"Observing you, my friend."

"Why?"

"Because you're almost there."

"Almost where?" Jason asked.

"At the point of understanding. You can't beat old Cris Croc alone. Remember the fifth point on the star?"

Jason thought for a brief moment. "Yes, it's 'Evaluate Results.'"

"Can you evaluate this hunt alone?" Alex asked.

"No, the whole team will provide feedback."

"So will your measures of success. They provide even more important feedback. Keep using them. Don't lose the ground you've gained by not tracking what you've already accomplished."

Again, Jason hesitated. "I've got to ask you, Alex: why do you keep calling King Croc a problem?"

"Because that's what he is, Jason," Alex replied. "Look, being optimistic and confident is great for morale – your own and that of your team – and I'm all for it. But there's a difference between confidence and a stubborn refusal to face reality. If you think of King Croc as only a challenge, you won't understand how difficult he will be to defeat. Worst of all, you're being less than honest with yourself and your team."

He paused a moment and then continued, "There comes a time, Jason, when you have to call a problem a problem. There's no shame or defeat in that. It's no reflection on your competence as a manager or the stability of your company or anything like that. Problems do come up – it's just a part of the normal cost of doing business, if you will – but you have to recognize them as problems before you can do anything about them."

Jason was silent for a spell, and finally said, "All right. So I recognize ol' Cris Croc as a problem… what then?"

"Evaluate both your results and your actions. Do another round of 'Lessons Learned.' You'll need to identify root causes, develop new approaches, and test them before your next hunt. Your team will tell you what needs to be done."

"You're still reminding me that I can't do it all myself."

"I am."

"I understand," Jason said.

"Almost, but not quite. You are, however, headed in the right direction. Tonight, you rewarded your team – not just with money, but with recognition of a truly collaborative effort. Build on that feeling. And most of all, don't *lose* it."

"Thanks, Alex."

"You're welcome… and *that* was no problem."

The two men walked back to camp in silence; there was no more that needed to be said. Jason fell onto his cot and slept at last – a deep and untroubled sleep. He awoke to the sounds of his team breaking down the camp for the return journey.

Jason's Reflections

The return to Supply City was uneventful. The fact that only a few supplies remained upon their return attested to their having planned well, and the team had recovered the old Jeeps and even the personnel carrier, for which, as it turned out, a grasslands hunter was a ready buyer. Jason was glad to be rid of the old thing; it was a burden and, for him, absolutely useless.

The bankers and financiers were justifiably overjoyed by the preliminary reports Jason had phoned in; this hunt had vastly exceeded their expectations. Several buyers had already bid for the gold and silver that had been recovered. Jason had not actually thought about the value of the silver, but it served to significantly increase the overall return on investment.

Tomorrow, they would begin breaking down the base camp in Supply City, and the following day they would be ready to return home. Everyone, including Jason, was ready to plan the next hunt, although it would not begin for another six months.

Alex left on the afternoon of the day they returned to Supply City. He had work that remained unfinished, but Jason and his team planned to meet with him at the first planning session in four weeks. For now, though, Jason and his team had to complete their "Lessons Learned" and begin evaluating the two hunts.

Stomp Yer Croc!

As evening fell, the Senior Safari Lead was organizing the breakdown of the camp. This gave Jason the chance once again to climb into a Jeep and drive away from the base camp. He stopped at one of his favorite spots, the hill looking toward the road into the Valley of Gold.

It was another one of those pristine nights, with the stars flung helter-skelter across the heavens. There was a slight whisper of a breeze that felt refreshing against Jason's face and arms and legs. Away from the bustle of the camp, the sense of calm was remarkable, and Jason suddenly noticed that he hadn't felt this good in a long time. Part of him wanted to just sit back and enjoy the feeling, but he found that his thoughts were still whirling.

We know how to do things differently now, he thought to himself, smiling. This thought was immediately followed by a vision of King Croc – a topic that was never far from his conscious mind – but the familiar jolt of disappointment, the stab of frustration, did not accompany the vision this time. Now Jason's thoughts were calm and reasoned.

"I only wish I'd recognized King Croc earlier," he said aloud. "If we had figured out how to deal with him, this hunt would have been more successful – maybe a lot more successful."

Jason's thoughts turned to how different each of his challenges – no, they were problems – had been. Pulling a note pad from his satchel, Jason started to write:

DESK NOTES
KNOW THE MISSION FIRST!!!
✓ Use planning that makes sense
✓ Make a commitment - - - stay involved and follow through
✓ Analyze the problem
✓ Use teams to find alternative solutions
✓ Test the best solution
✓ Measure the results - - - <u>and don't stop measuring</u>
✓ Learn from your successes and not just your failures

He stopped writing and re-read the list. "I know what's missing," he said aloud, and started to write again.

"I want the team to keep me straight on these," he mumbled. "Maybe I should have them look at this list. I need to know what they think."

He rested the note pad on his lap and continued to reflect on the hunt. His last conversation with Alex at the crest of the ridge came to mind. *Alex is right*, he thought, not for the first time. *We're solving real problems, not merely overcoming challenges. But when we solve a problem, what is it that we're really doing?*

Jason flipped over to another page and began to draw, this time, a four-pointed star.

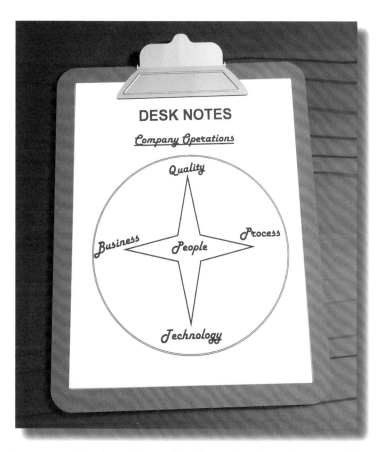

Once again he found himself talking aloud. "Solving problems is making improvements," he said. "When we make improvements, we can save money and make money at the same time."

Then he thought about King Croc again. *What kind of problem is he? Jason asked himself. Is he a unique problem unto himself, or is he an amalgam of the different problems we faced, grown to supernatural size by our own lack of understanding?*

As he drifted off to sleep under the stars, Jason wondered if old Cris Croc wasn't an entirely different kind of beast. *Maybe*, he thought sleepily, *Alex will help us. Then, again, my team may find the answers, too.*

However, all of those thoughts, and all of those questions still waiting for the right answers, would wait until morning. For now, Jason drifted into peaceful, contented sleep such as he had not known in a long, long time.

Stomp Yer Croc!

the post-game meeting

The day after the game, Carson and his team met at lunch to talk about the story of Jason and the crocs. Once again, Carson set up his laptop and the projector and set pads and pens at each place at the table, knowing they'd need to take some notes. Buddy and Coco walked in together from a budget meeting. As usual, Marsha was the last to arrive.

"Is everybody ready?" Carson asked.

Coco, the biggest football fan in the group, said, "I'm still buzzing over that game last night. We blew them out. I couldn't believe that safety in the last minute."

"Yeah, 29 to zippola," Buddy exclaimed. "Incredible win, and it puts us in the playoffs for sure."

Carson was less impressed. "We still need two more wins to guarantee the playoffs. You know what they say about any given Sunday."

Buddy scowled. "Right. And none of our next competitors has a winning record."

Not to be excluded, Marsha jumped in. "We didn't need another injury in the secondary. Two of those next opponents are big-time passing teams."

"We need to start this meeting," Carson said. "I unplugged the phone in here. Everybody have their cell phones on stun?"

Everyone said yes. "Buddy, are you sure?" Coco asked. "Yours rang twice in the last meeting."

Buddy re-checked his phone. "Yes, I'm sure. Let's start."

"All right," Carson said. "I guess the best way to begin is for all of you to give me your thoughts on the croc story. Just off the top of your head – doesn't have to be eloquent. What stuck out for you?"

Marsha wanted to speak first. "In the beginning, I was wondering 'What's with the ox carts?' Then, I thought about their environment."

Buddy was puzzled. "What environment?"

107

"Jason's team was used to hunting on plains and in the grasslands," she replied. "The ox carts are downright medieval, but Jeeps get stuck in the mud. An ox cart can go through almost anything."

"They're also slow," Buddy shot back. "Jason's team made better time when they got rid of the ox carts."

"Okay, Marsha," Carson interrupted, "how many ox carts do *we* have?"

"Now, you're being ridiculous," Coco replied. "Budget Com Services is a high-tech financial services company. When – or even why – would we use an ox cart?"

Carson looked at her, one eyebrow raised, not knowing whether to believe her or not. "C'mon, Coco. You know we're not talking about real ox carts. What was the story really about?"

"That's the point, Car," Buddy said. "We all got it."

Coco nodded as Buddy continued. "Let's forget the ox carts and all the other metaphors and cut to the chase. In the end, it's just the same old process improvement tale. Besides, we've been there, done that at BCS. We got rid of our 'ox carts' long ago."

"Is it really just the 'same old process improvement tale'?" Carson asked. "Were processes the only thing Jason improved?"

"He got more gold," Marsha said a little impatiently. "But as Buddy said, our situation is different. As I'*ve* said on numerous occasions – not that anyone seems to be listening – we're already successful and profitable. Spending more money on process is a waste."

"Marsha, we do listen to you, and we value your input," Carson said, but Marsha didn't exactly look convinced. "But I'd like to point out that Jason Hunter's business was successful and profitable too."

"Agreed, but they were doing a lot of things in a less than optimal way – at least in the 'pre-Alex' days. For example, they'd make those elaborate plans before a hunt, but then they would rarely consult the plans. They were kind of haphazard in several ways," Coco said.

"Exactly!" Buddy said. "That doesn't describe our firm, not by any stretch." It was clear to Carson that his friends had really missed the point of the story. So he decided to try a different approach. "Did we win the game last night, Marsha?"

"You know we did. You were there."

"Did we score on every play?"

"Of course not, but what's that got to do with the story?"

"All of you bear with me," Carson said. "At the beginning of that seminar I was telling you about, I felt the same way you did, but by the end, I'd changed my mind. Just hear me out, okay?"

Coco was more open to the discussion than Marsha or Buddy, but they all agreed to give Carson a chance. After all, he was their trusted friend and partner. Unlike all too many business owners or managers, Carson wasn't the type to blindly embrace every new seminar that came along, just because it was the thing to do. He valued his company's money and his team's time too much to waste either. If he was so passionate about this croc-stomping stuff, he must have had a very good reason.

Carson continued, "We only gained 32 yards up the middle and they stopped us from completing any long passes. By going off tackle and using short passes, we took them apart."

"They almost scorched us twice with long passes, but we held them in the red zone. Their kicker missed an easy field goal, and we kept them out of range on the next drive," Coco added.

"Right," Carson agreed. "We didn't play a perfect game, but we still won. Now let's think about Jason. He succeeded on both hunts, yet neither hunt was perfect. BCS succeeds. Are we perfect?"

Buddy interjected, "No one – no team or company – is perfect."

Carson nodded. "That's why we always hear the mantra about 'continuous improvement.' The problem is that we think in terms of continuous process or quality improvement. It's a trap. That vision of improvement is too narrow."

"You're talking about that four-pointed star at the very end of the story," Coco said.

"Absolutely," Carson replied. "We need to think in terms of organizational improvement. Not all of the answers are in process or quality. Sometimes, there is waste or inefficiency that has nothing to do with either one."

Marsha remained skeptical. "We still won't be perfect."

"You're right," Carson agreed. "But remember, that's not what we're striving for. Besides, when you think about it, 'perfect' is a subjective judgment in some ways. Sure, we have various criteria to objectively measure how 'good' we are in many areas – but even if we were to score 100 % in those areas there would still be room for improvement in other areas that can't be measured so easily. For example, Marsha's idea of a 'perfect' company might be at odds in some ways with Buddy's. And so on."

Coco, Buddy, and Marsha nodded, waiting for Carson to continue.

"The point is that we can be better, and we can keep getting better. In addition, we all need to stop thinking about improvement as a cost. An improvement that doesn't save money isn't an improvement."

Coco laughed. "Now, you're in my world. Our focus is always on getting a return on investment."

"It's your world and should be everyone else's, as well. Folks, we're already ten minutes into this meeting, and we're just starting to make progress," Carson observed. "There's a lot of ground to cover."

"I'm beginning to see that," Marsha said. "But I'm wondering about this focus on the crocs. I mean, just what *are* the crocs? It wasn't a croc hunt; it was a hunt for gold. Still, the crocs seem to be the focus of the story."

"Crocs are problems of various kinds," Carson answered. "And like most companies, we have our share of them."

Marsha was defensive. "I don't like to say we have problems. It's too negative. We have challenges. If we say we have problems, it's looking at things too negatively."

"Well, that's what Jason thought, too," Buddy pointed out. "That is, until Alex got him to treat the crocs as what they really were."

"Exactly," Coco added. "I think most of us are so indoctrinated with 'positive thinking' – or, more recently, that silly 'law of attraction' fad – that we're actually afraid to call a problem a problem. Or maybe it's just a matter of pride or saving face or whatever you want to call it. It's almost as if we're all so focused on presenting a perfect picture to the world that we're not really being honest with ourselves."

"I have to admit you've got a point there, Coco," Marsha said.

She continued, "Maybe we need to leave the spin to the folks in corporate communications and focus more on looking honestly at our problems. Even if we have to be brutally honest at times."

"Now you're getting it!" Carson said, smiling. "Think of it this way: How much priority do challenges get? Until they escalate to crisis level, we ignore them or give them lip service. Prior to that point, they cost us money. If we call a problem a problem at the beginning, it gets our attention."

"But the problems Jason and his team had weren't all crocodiles," Coco remarked.

"Yeah," Buddy exclaimed, "There were the ox carts, monkeys, SUVs, untrained people, and even people like Noel. He reminded me of Tim in Engineering or George up in Finance. They're so set in their ways. Don't forget about the roadblocks and the wall, either." There was a general murmur of agreement in the room; they all understood these symbols.

Car started to relax. Now things were beginning to sound more like they did in the seminar. "You're right if you're thinking that all of the people and animals and objects in the story represent real situations in the business world. Tell you what. I have some slides from the seminar. Let's do the crocs first. Then we'll take on some of the others."

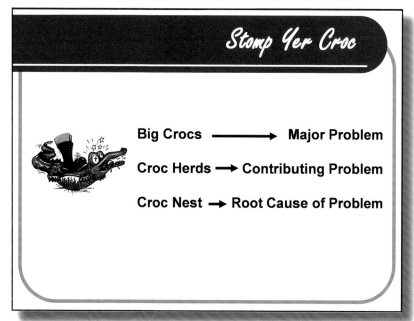

As Carson had promised, the first slide addressed the croc problems. Coco had the first question. "What about the difference between swamp crocs versus valley crocs?"

"Let me answer your question with another question," Carson replied. "What did the Swamp Crocs do?"

"Interfered with the journey to the valley," Marsha offered.

"The valley crocs were different because they kept the hunters from getting all of the gold," Coco added.

"So… what are the equivalent 'valley crocs' and 'swamp crocs' at BCS?" Marsha prompted.

"One way to relate them to our organization is that swamp crocs are technical and project problems," Carson said. "Valley crocs could be our organizational problems. But there are other ways to look at these two croc types as well – ways that might be even more useful for our purposes. Anyone have any ideas in this regard?"

Coco, who had appeared to be deep in thought for a few moments, responded, "Well… maybe we should consider the different locations of the croc types."

"What do you mean?" Buddy asked.

"Let me think about this a moment," Coco said. Then, shaking her head, she added, "Never mind… Go ahead, Car."

"No problem. If you remember from the story, swamp crocs are closer to their nests," Carson began, "and the nests are the root causes of problems. In this context, swamp crocs could represent more apparent problems whose root causes are more easily identified."

"The valley crocs also located their nests farther away," Marsha said, jumping back into the conversation. "Does that mean they are less obvious problems?"

Car grinned. "You got it! The problems appear to be the same on the surface, but dealing with them requires a different approach. The root causes of these problems are harder to find and more difficult to resolve. They also tend to be problems that we might take for granted as just being part of doing business, which is why we don't see what they cost us every day."

Buddy finally asked the obvious, "Then exactly what, or who, is King Croc?"

Carson smiled broadly. "I would rather save him for later. He's a little more complex to understand until we talk about some other parts of the story."

"Let's take a run at it anyway, Car," Buddy said.

"Well, all right. King Croc is a huge menace and a major drain on our treasury," Carson explained. "The valley crocs are with King Croc because they're the problems we fail to see. We don't recognize them as problems because they are institutionalized as a part of the way we do business. King Croc is the epitome of all of these fearsome creatures. As I said, we'll go into more detail about him later."

"Okay, I understand," said Buddy. "Let's talk about the monkeys. I think they're just what everyone called them, namely distractions."

"Then, if you'll excuse the pun, Minkee the Monkey is about pet projects," Coco ventured.

"Which may or may not be problems, per se," Marsha said. "If you remember, Car, our best product line came out of Hank's pet project."

"True enough," Carson replied. "It's how we foster them that really makes the difference. That's every bit as important as getting rid of the bad ones before they cost us time or money."

"Hey," Buddy interrupted, "How come no one has mentioned the chickens yet? They were an interesting twist."

"I'd have barbecued the lot of them," Coco said. "But, then, I'd barbecue anything."

Everyone laughed at Coco's comment, and the conversation turned to reminiscences about her culinary skills, memorable instances at their get-togethers over the years, and plans for the next company barbecue.

After a few moments, however, Carson decided he'd better get the group's focus back on topic.

He interrupted the back and forth, saying, "Okay, folks. It's obvious that even with full bellies, we tend to obsess about food, but we really need to keep going, or we'll never get done."

"If we must, Car," Coco responded. " The subject was chunky chickens, I believe."

There was yet another round of laughter, followed by everyone returning their attention to Carson.

"Chickens it is," Carson answered. "Here's a slide that may help."

Stomp Yer Croc!

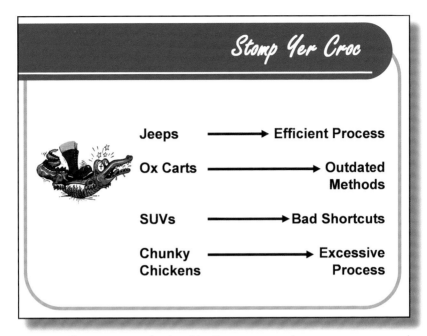

"Actually, come to think of it, I'm not sure how an ox cart and a chicken are different," Marsha sighed. "In the context of the story, both seemed pretty useless. Although the chickens did provide food for the team at one point."

"It's actually quite simple," Carson said. "A chicken... or, as in this story, a chunky chicken – is a bloated or overly-complex process. It slows us down by making us take the time to account for things that probably don't need to be done."

"Okay, so what about the ox carts?" Marsha asked.

"Wait, I'm coming to that. The ox cart can, however, represent the practice of following laws or regulations, or responding to customer needs, that no longer apply. They were useful in their day, but that day passed a long time ago."

"What's wrong with SUVs?" Buddy challenged. "After all, they were late-model, high-end vehicles. Seems to me that good technology is something we should use."

Carson had heard that comment at the seminar. "It's good only when it's the right technology for the right solution. The SUVs had great features, but they weren't well suited to the swamp hunt. As you see on the slide here, the SUVs could represent poorly thought-out shortcuts and other mistakes. You know the old saying, 'There's never time to do it right, but there's always time to do it over.' Sometimes the shortcuts we take end up costing us more in time and money than they're worth."

"You know," Marsha mused, "Now that I'm thinking of it, this all seems like an overly complex way to describe problems. Aren't they all still crocs?"

"It wasn't really that simple, if you recall," Carson replied. "The crocs were one category of problems Jason and the team faced, but not the only category. If you noticed on the second hunt, there were fewer crocs. That's because they took care of some of their problems by eliminating their root cause. I think the lesson is that process problems are just one type of organizational concern – but not the only type."

"My guess is that we make our own problems by choosing the wrong kind of processes," Marsha said. "Inefficient processes are a type of problem…"

"But not the only type," Coco interjected. "And if we don't recognize them as problems, they continue to cost us money and time."

"Sounds like any kind of problem costs us money," Marsha said. "The question is – how much?"

Coco responded before Carson could reply. "How much it costs to get it fixed ought to be looked at, too. If it costs more to fix it than to leave it alone, I'd leave it alone."

"We also need to look at whether it stays fixed, and more than that, how we know it is staying fixed," Carson added, "but we'll cover that a little later."

"The *chevaux-de-frise* puzzled me," Buddy said.

"What do you mean?" Carson asked.

"Well, do they represent a real problem, or some kind of obstacle or barrier? Do you have a slide, Car, or are you just going to tell us?"

Carson said with a smile, "I have a slide. We'll probably talk some more about this one as well." He switched to a slide that listed three items: "Roadblock," "Wall," and "*Chevaux-de-frise*."

"The roadblock isn't really such a big deal," Carson continued. "We plan for them and deal with them every day. In the tale, there were just two walls, and both led to dead ends."

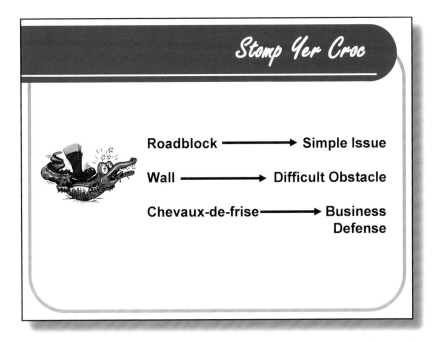

"But there was gold behind the wall in the valley," Coco objected.

"Yes," Carson replied, "and that's because Jason and his team hadn't figured out how to get by King Croc."

"So… maybe King Croc represents problems that we don't want to deal with because they invade our comfort zone?" Marsha asked.

"To paraphrase Alex, give that lady a cigar!" Buddy said with a very mischievous grin.

Pretending to be insulted, Marsha retorted, "I don't smoke, thank you."

"Why are the *chevaux-de-frise* in with the barriers? Don't they protect us rather than just keep us back?" Buddy asked.

"That's a good question," Carson replied. "Here's my take on it. Sometimes they do afford us protection when they give us time to react or re-plan. On the other hand, if we use them to tune out our customers, or messages that we don't want to hear – whether they come from the employees or from outside the company – they quickly become counter-productive."

"Am I right to say that communications can act as a sort of *chevaux-de-frise*?" Marsha asked. "It isn't good business diplomacy or good PR to be *too* honest about a company's weaknesses."

"Are you still advocating for the word 'challenges' instead of 'problems'?" Carson responded.

"Not really. I'm just not sure about how candid a company should be with its employees – or the public. It's one thing for the management team to be honest with themselves and each other about calling a problem a problem. It's another thing to broadcast 'problems' to the whole work force, the customer base, and the public."

Buddy jumped in, "Yet if there's something going on… and if the staff doesn't know for sure just what's happening, rumors start to fly. Before you know it they take on a life of their own. That's not good for a company either."

Marsha sighed. "I suppose, but… well, maybe we need to leave that for a more in-depth discussion later. To me it still leaves the issue of how we make the needed changes, even if we recognize the problems."

"You sure do like to jump around from subject to subject," Buddy said teasingly, but he was glad she had moved on. He thought Marsha had a tendency to take things too seriously. He'd often told her, "You think too much."

"Actually," Carson said, "Marsha raises a very good question. In fact, that was pretty much the point of the seminar." Marsha stuck out her tongue at Buddy as if they were school kids.

"Teacher's pet," Buddy mumbled good-naturedly.

"Okay, kids, if you don't pay attention now, there'll be no recess," Carson chuckled. "If I may be allowed to continue… If you remember the five-pointed star introduced earlier in the story, it defines the overall approach. The tale provides some other clues."

"I think we're already doing everything in that approach," Buddy replied. "Our certifications require them in one form or another."

"Those certifications don't save us a penny," Coco said. "We only have them because they're required in our business, and we do what we have to do so we can stay in business. Car is saying that BCS can actually be more profitable or something."

Stomp Yer Croc!

"It's more of the 'or something,' but profit is a part of it," Carson responded. "Redirecting effort – which we ultimately measure in dollars – may not directly result in an increase in profit. It may result in using our resources more effectively to get more done."

"Okay, so we can lower costs or do more with what we have. Is that a better way of saying it?" Coco asked.

Carson nodded in agreement. "Let's look at that first starpoint, because the presenter had some important comments." Carson put up another slide. This one was labeled, "Commitment," and had several bullet points, the first of which was "Vision, Mission & Core Values."

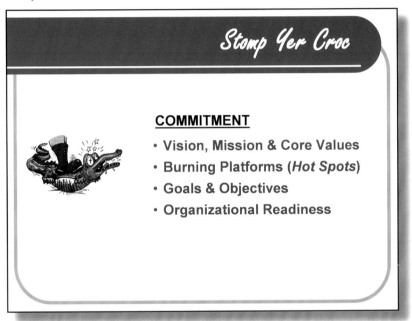

"We re-wrote our mission and goals last year. How do we know that we achieved them?" Carson asked.

A puzzled look came across Marsha's face. "We exceeded our sales targets, we stayed under budget, our market share increased, and the stock value went up."

"What about our per-unit cost? What happened to our staff productivity and turnover rates?" Coco asked with a grin.

"Those are chal-leng-es..." Marsha said as her voice trailed off.

"Those were Coco's 'hot spots,'" Carson said, pointing to the second bulleted item on the slide, "but their importance was minimized. How much better could BCS perform if those issues were improved?"

"I can't take on these issues alone," Coco added. "I'd need help from Buddy and Car to work with my people just to find out what the contributing problems are."

"Those are two big crocs, and we only see part of their croc herds," Buddy put in. "We have no idea about the nests."

Carson started to get excited. "That's what this first star point is about! It's focused on just getting committed to solving problems – and then organizing to make change happen."

The third item on the bulleted list was "Goals & Objectives." Once again Marsha had a question.

"Aren't goals and objectives two words for the same thing?" she asked.

"Aren't goals and objectives two words for the same thing?"

"No, not necessarily," Carson answered. "Many people use the two words interchangeably, but they're actually two different things. Goals are broad, whereas objectives are narrower. To put it another way, goals are general intentions, and objectives are precise landmarks."

"Or… goals are abstract and objectives are concrete?" Coco offered.

"Up to a point, that might be a good way of putting it," Carson replied. "This seminar emphasized that are two kinds of goals. Strategic goals are long term and are where we want to be three to five years in the future."

Marsha interrupted, "I'd guess they're tied to our Vision and Mission."

"Absolutely," Carson said with a smile, "and we build that future by attaining the other type of goal – our annual, or tactical, goals, which aren't all that abstract. Our annual goals have some high level targets. We validate the attainment of annual goals by verifying the achievement of the objectives that go with them."

"So objectives are more precise?" Coco asked.

"Precise, but not rigid," Carson answered. "Depending on how complex the goal is and how many objectives are necessary, you won't always accomplish 100 percent of the objectives. But I think a good rule of thumb is that if 80 percent of a goal's objectives are achieved, then that goal has been reached. Objectives are the measures we use to validate when and how a tactical goal is attained."

"I'm not sure I like being even that accountable," Buddy said. "You know, sometimes stuff just happens. For whatever reason, objectives don't get met."

"Yeah. Remember the days of 'zero defects'?" Coco said with some contempt. "How long did that last?"

Everyone in the room chuckled aloud.

"Got that right, but that's okay," Carson assured Buddy. "The teams responsible for an objective just roll over what didn't get done from one year into the next. Not meeting an objective isn't a crime. As long as 80 percent or more of an objective is achieved, the effort is successful. We should always keep our eye on the goal, and sometimes even adjust it

– and the objectives – as needed. As a team, we should be helping each other and not beating up on folks."

"Got it," Coco said. "But how do we decide what our goals should be, let alone determine the objectives necessary to reach those goals? I've been to department meetings where people got into some pretty heated arguments about whether their top goal for the next quarter should be a ten percent or twelve point five percent increase in sales… stuff like that. It just seems like such a waste of time, and it's bad for morale too."

"No one should argue about a goal," Buddy agreed. "It should reflect an obvious problem or concern." Carson nodded.

"I think it bears repeating that if we've done everything we can, then not meeting an objective isn't the end of the world," Coco said. "I know people who beat themselves up for falling short a couple of percentage points or whatever, instead of focusing on what they did right, and on how they can do it better next time. From my point of view, that's a real waste – it's counterproductive, and it's a real bummer. I mean, just because they call it 'work,' doesn't mean it can't be enjoyable."

"Precisely," Carson said, and his smile was broader than ever. "That's really another thing that this seminar was about – making improvements and actually having fun at the same time," he said. "Improving a business is a never-ending process – or it should be, anyway. And we're never going to be 'perfect' by anyone's standards, but so what? As long as we have a steeper rate of improvement, we're a success."

"So it's not just about improving, but also *how* we improve?" Buddy asked.

"No," Carson replied. "By focusing initiatives across the organization and doing a few things well instead of doing a lot of things with modest success, we'll gain that steeper rate. If everyone is having a good time while we're achieving significant improvements faster, then I'd say we're a super success."

Now, Marsha had caught Carson's enthusiasm. "I see your point. As long as we make progress, we succeed. Progressive improvement and not perfection is the goal."

Stomp Yer Croc!

ESTABLISH DIRECTION

- Prioritized Problems
- ROI Expectations
- Specialized Team Assignments

The next slide flashed onto the screen: "Establish Direction."

"On the first go round, the objectives we establish are pretty loose. They get tightened up a bit in the second star point," Carson said. "Part of getting organized is getting help. We set the strategic direction, but our managers are more involved in the actual processes, and can provide tactical details that we may not consider or even be aware of."

Pausing to let them consider this, he continued. "In order to coordinate our goals with the objectives required to reach them, we'll need an interface – an improvement group – to support the teams."

"How good do those ROI estimates need to be?" Coco asked, pointing to the second item on the bulleted list. "Seems to me that we wouldn't have enough data yet, so early in the process."

"That's a good question," Carson acknowledged. "To answer it, those estimates have to be good enough to prioritize the first group of big problems."

"Not challenges… problems," Marsha said, smiling.

"Right you are," Carson continued. "Actually, it's a pretty simple process. The current operational cost is compared to a vision of the optimal future cost. Then, the total cost to eliminate identified problems is subtracted."

Coco seemed satisfied but had some reservations. "Why not do a more detailed analysis at the beginning?"

"We wouldn't know enough," Buddy replied.

"Besides," Carson interrupted, "all of the problems can't be solved at once. The prioritization of problems helps determine which problems get solved first."

"What are the selection criteria?" Marsha asked.

Carson didn't want to be too precise. "It depends. Those problems that are easy and cheap to fix might be the first chosen to show some quick success. We could also pick a few with higher ROI to get some big savings early."

The other three executives exchanged approving glances. "I have several big problems to be considered once we decide to go forward," Buddy said.

"Once we decide on the problems," Carson added, "the managers can put together teams with the expertise to do more detailed analysis. Buddy, if your problems are on the list, you can help put together the teams."

"I want to hear more," Marsha said. "I'm interested, but I want to understand how this is different from what we're already doing. I think all of us still have a bitter taste in our mouths from the 'improvements' we made in response to that consultant we hired a couple of years ago. If you recall, Car, it took us two full quarters to even get back to where we were before he supposedly helped us."

Coco chimed in, "I want to hear more, too. It actually took my people closer to six months to untangle that mess"

"Add me to the list," Buddy said. "This sounds good, but, like Marsha and Coco, I'm also thinking it sounds pretty familiar. We all want to make things better around here, but we sure don't want to 'improve' our operation like that again."

Carson knew they were almost convinced because they were finally engaged. He put up the next slide, labeled, "Define Success."

Stomp Yer Croc

- **DEFINE SUCCESS**
  - Resource Requirements
  - Measurable Objectives
  - Plans & Deliverables

Coco started to object to the idea of "Measurable Objectives" being set so late in the approach. Realizing the roots of her objection, however, she asked, "When we refer to 'Measurable Objectives,' what kind of measures are we talking about?"

"The teams will uncover them based on their research. Each 'Measurable Objective" is going to be a unique function of the specific problem being addressed. Engineering could focus on aspects of defect tracking, cycle time, or estimation accuracy. Finance, human resources, and IT each would establish measures that prove the problem is eliminated," Carson said.

Marsha interrupted, "I'm more concerned about an explosion of plans. More wasted time on writing shelfware – kind of like the elaborate plans Jason and his team created, but never actually referred to during the hunt."

"Can't blame you there," Buddy added. "We've all seen that happen too many times."

Carson wanted to overcome this objection quickly. It had unnecessarily consumed a lot of time at the seminar, and he didn't want the same thing to happen here. "No, we don't need to make a bunch of huge plans. We'll only use charters for the team, and they'll be brief and to the point. The overall improvement approach is to always apply the old K.I.S.S. principle."

"Like we do with Six Sigma," Marsha said, and added, "Does this approach use Six Sigma?"

Carson liked this question, as it addressed the real strength of the approach he was describing. "We can use any tools that we think will work. The approach is independent of any particular method."

"I'm still not seeing any one thing that differentiates this plan from so many others," Marsha said.

"We're coming up on several things," Carson assured her. "First, it's different because this is a long-term approach with big short-term returns."

Coco added, "It's also going to take the best of other methods and techniques. That's pretty clear. Personally, I like the evolving analysis of the return on investment."

"As you'll see in a moment, this approach keeps on measuring results to ensure that we hold the ground we've gained," Carson said.

"Like Jason and his team did in the valley on the second hunt," Buddy suggested.

Carson wanted to move on. "Yes, that's correct. Now, let's get on to the last two slides. Here's the fourth star point: 'Deploy Improvements.'"

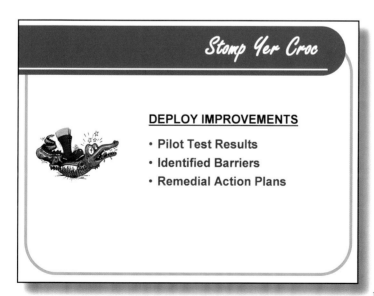

As Carson brought up the next slide, he could almost see the light bulbs clicking on above everyone's heads. He smiled to himself, thinking, *They're really getting it! This is turning out to be less of a tough sell than I thought it would be.*

Carson said, "We'll run a small-scale pilot test of the solutions we come up with, and figure out where they are working and not working. This allows us to correct errors before rolling it out to the entire organization." He gave the group a moment to look at the slide before he continued.

"This is the point at which the biggest difference begins. We've implemented organization-wide improvements based on solutions derived from a causal analysis. We will have built the measures that tell us when a solution is working. Where we've failed in the past is in not continuing to use those measures to evaluate our ongoing efforts."

"Now, I see what's been happening and why you're excited about this," Marsha observed.

Buddy and Coco nodded in agreement, and Coco added, "As a result, each department can continue to track their success so that we save money year after year."

"We can build on our successes by addressing additional problems each year, using the same approach," Carson said. "By using the same approach, but continually updating the information we have, the successes we realize will be an ongoing flow, rather than individual spurts of improvement."

"The last slide I have relates to the fifth star point, he added. It's focused on evaluating the results of our improvement efforts."

As he walked back toward the easel, Carson couldn't help but sense the excitement around the table. Smiling to himself, he felt that the entire group was more energized than he had seen in some time.

Finally, with a bit of theatrical flair (and a bit of well-earned self congratulation for having given what he felt was an effective presentation), he flipped the page to the last slide, titled, "Evaluate Results."

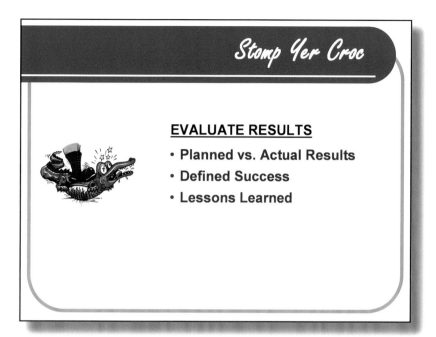

Coco saw the slide and burst out, "We'll know how well we succeeded based on the numbers. This ties back to the objectives and goal attainment."

"The whole process is completely self-contained," Buddy added, "just like a well-defined network."

"And that's it," Carson said. "The only question is whether or not we try it out."

The answer to Carson's question was evident in the excitement on his leadership team's faces. They were, to a person, convinced.

A New Beginning

The next day, Marsha enthusiastically called the consultant herself. A kickoff meeting was scheduled for two weeks from that date.

The two weeks flew by; there were dozens of preparations to be made. Almost before she knew it, Marsha was walking with Carson into the large meeting room, where the workers were taking their seats and preparing for the meeting to begin.

"We have a clear mission, and our people are about to give us some challenging goals," Carson said to Marsha.

Stomp Yer Croc!

"Actually," Marsha replied, "I think there will be some interesting crocs to get stomped."

Carson smiled. "I think I can safely say you're right about that."

"In fact," Marsha continued, "I want to find our King Croc. My team is ready, and I've given them some healthy incentives."

"I want to find King Croc," Marsha said. "My team is ready, and I've given them some healthy incentives."

There was a flurry of last-minute preparations, the usual round of introductions, and then the consultant took the stage and began.

"I have five words for all of you," he said. "Be ready for an adventure. This isn't going to be just another boring workshop or seminar or class. We are going to have some FUN here – identifying problems, finding solutions, and – most of all – creating substantive, measurable, positive change. It's all up to you. Finally, I have just one question for you."

He paused and looked around the room. "Are you all ready to STOMP YER CROCS?"

A loud cheer erupted. Carson, Marsha, Coco, and Buddy had big grins on their faces as they looked at each other and then around the room.

And so the adventure began…

What are you going to do . . .

. . . about YOUR crocs?

Stomp Yer Croc!

some final thoughts

I have traveled the world over the past twenty years, from China and Europe to Australia and the U.S. I have worked with small, medium, and large organizations in virtually every industry, in both the public and private sector. And I have found that at least 80% of the problems my clients face are similar, regardless of the industry, the size of the firm, or whether it is a public or private enterprise.

What I have seen is that organizations are doing many good things, but it is all too easy to lose overall focus as a company. As a result, the company's basic systems become less than optimal, very often without the full knowledge of anyone in the company. There may be a general sense that something is "wrong" or that it could be better, but there's very little insight into what the problems are. In fact, many managers and CEOs are reluctant to even acknowledge that problems might exist, preferring to call them "challenges" or some other euphemism.

Problems do exist, even in the best-run companies. But they're not the end of the world, and they're not necessarily a negative reflection on the company. They certainly don't signify failure. In fact, the issue is not that organizations fail to be successful or even profitable; rather, they sometimes struggle with fully achieving the following:

- Optimal efficiency across the organization
- Reasonable effectiveness from their employees
- Mutually supportive interactions among people, process, and technology
- Consistent customer satisfaction from higher quality products and/or services

As a result, countless companies are simply not as successful or profitable as they could be. In some cases, they are dramatically *less* successful than they could be if they were more aware of the "crocs" and other problems in the organization, and were better able to deal with them.

Stomp Yer Croc!

If you've read this far, you know that "stomping crocs" is not merely a matter of putting out fires or taking care of day-to-day "situations" in order to keep your head above water. It's not even just about solving problems – yes, there's that "p" word again. There is always a larger goal – as Jason Hunter found out, there is almost always more "gold" to be discovered and obtained.

This is not to deny that existing methods and standards are having a positive impact on industry. They are. Yet there remains a genuine uneasiness in both the private and public sectors as to whether more can be achieved.

When I talk about "achieving more," I hope you know that I am *not* talking about plain old-fashioned greed, for which business, particularly big business, has taken so much flak over the years (some of it justified). Rather, I'm referring to the genuine desire that good people have to improve their way of doing business, with the end result being better products, more satisfied customers, and happier employees and shareholders. The "gold" to be discovered is certainly literal in many cases, and profitability is a big part of the picture, to be sure (as we will discuss below) – but it's not the *only* factor.

I have seen a deep desire for improvement again and again when speaking with business owners everywhere, from Schenectady to Shanghai, from San Francisco to Sidney. Call it "divine discontent," if you will, a persistent feeling that something better, maybe even *much* better, is possible for these companies. I believe that such improvement is not only possible, it is *doable*. And my belief comes not from abstract theorizing, but from having seen firsthand how much difference a superlative method, properly taught and implemented, can make in all of a company's functions.

Positive Results Are Real!

Over the past 15 years, Carnegie Mellon University's Software Engineering Institute (SEI) has reported significant savings as a result of CMM and CMMI®-driven process improvements. For example, the SEI recently reported that over a six-year period, one user of CMM® and CMMI® realized a productivity improvement of 76%, resulting in a cost savings of $412 million.

Ralph Williams

The issues raised in *Stomp Yer Croc™* are not that process improvement and quality improvement fail to produce measurable results. SEI has historically reported data on several key performance indicators on those companies have implemented CMMI® that demonstrate what can potentially be obtained through an organization's process improvement efforts alone.

Annual Improvement Results	SEI Average	SEI Best
Increased Productivity	62%	255%
Cost	20%	87%
Schedule	37%	90%
Quality	50%	132%
Return on Investment	4.7:1	27.7:1
Customer Satisfaction	14%	55%

(Reference http://www.sei.cmu.edu/cmmi/results.html)

Numerous other studies over more than a decade have verified that process improvement increases productivity by at least 10% and as much as two-fold, if done right. Using process and quality improvement methods and standards alone, your organization should see these kinds of results. If you are not getting these kinds of results, you need to ask yourself, "Why not?"

As remarkable as these improvements are, however, they address only process and quality. What about the other kinds of waste and inefficiency identified by Shewhart, Feigenbaum, Deming, and Juran? The Star OI™ method outlined in *Stomp Yer Croc™* aims at the crux of these other problems, *including* process and quality concerns.

The phrase, "Show me the money!" was showcased in the movie *Jerry Maguire*. From an executive's point of view – a point of view that will get no argument from me – that is the key measure. Depending on the company, I have seen 20 to 40 percent of revenue saved when a company seriously applies Star OI™ style methods. When a company "holds its ground," as Jason did on his second hunt, cumulative savings grow from year to year.

133

Dealing With The Business Climate

One of my customers recently stated that there are so many models and standards available that he struggled with deciding which one to use, or whether it would be better for him to attempt to use two standards and try to make them work together. He knew that improvements were needed, but was at a loss as to which model or standard, if any, would provide long-lasting results. I have also had clients in the early stages of their business improvement initiatives state one or more of the following objections:

- "I can't afford to spend time on administrative stuff like planning... we just go at it."
- "It's the nature of the business – we will always encounter operational problems."
- "We can't afford to install changes. We just improvise as needed. When the customer says *Jump*; we say H*ow high*?"
- "Even if we wanted to improve, where do we start?"
- "If we try to adhere to multiple standards, it will involve duplicated effort and wasted time. How do we integrate them effectively?"

Most, if not all, of the models and standards also require specialized training, in addition to the interpretation of principles to apply them to an individual organization. The materials are often too dry to secure solid employee involvement. Change is difficult enough for most people, even if it's a positive or welcomed change. Why would anyone want to change if the activity is dull or painful?

As a matter of fact, resistance to change is one of the biggest perceived issues regarding any kind of improvement program. The other major perceived issue is cost. I have found, however, that by showing companies their potential return with Star OI™, the cost issue typically disappears. By making change pleasant – even *fun* – by building real organization-wide rewards into the "hunt" – resistance to change actually dissipates over time.

The Institutionalization Challenge

If we are to take Shewhart, Feigenbaum, Deming, and Juran seriously, for every $100 million in annual revenue, between $25 million and $40 million is some kind of waste, loss, or inefficiency. Those are some pretty large – and scary – numbers. Process and quality improvement can account for the elimination of one-third to one-half of these amounts.

The unhappy truth, however, is that up to $26 million in waste, loss, or inefficiency per $100 million in revenue remains unaddressed. Further, process and quality improvements tend to be project-specific rather than organization-wide. The reasons for this are pretty easily identified.

Historically, process and quality improvements have focused on engineering and manufacturing, which are engaged in projects. The quality assurance and control, configuration management, and supply chain management groups are involved because of their project or product line-specific roles.

The result is that other operational functions, such as human resources, contracts, supply chain management, business development, finance, proposal management, facilities, safety, and security are largely ignored in the improvement efforts, or are addressed only partially. This reality only serves to support the general perception that process and quality improvement are "an engineering or manufacturing thing."

When the business goal of gaining compliance with a standard is focused on those projects that will be appraised or assessed, projects that are outside the scope of the improvement effort often are ignored within the company. This reinforces the challenge of institutionalizing broadly based improvements and change.

As great as the successes of process and quality improvement methods and standards have been, even greater successes remain to be achieved. That is what led to the creation of the Star OI™. Star OI™ was created specifically to deal with the business environment as it exists – including the institutionalization challenge.

A New Alternative

Stated simply, and from the "people" point of view, there is too much work and not enough enjoyment in making the necessary changes required to eliminate the rework, waste, inefficiency, and resulting loss that exist in organizations. It was in this context that the Star OI™ method outlined in S*tomp Yer Croc*™ was created. But I have always felt that some of the most effective ways to get a point across are through storytelling and illustration, rather than the endless recitation of principles and dry statistics. That's why I came up with the S*tomp Yer Croc*™ story in the first place, enhancing it with color cartoons and, hopefully, amusing characters, to make my points. The fable and "fun" elements of the book reflect the nature of the Star OI™ method itself. My hope is that this tale serves my intent to make change desirable, enjoyable, and rewarding for companies everywhere.

Star OI™ is a structured and question-based method – a hunt, if you will – that improves businesses quickly, systematically, and effectively. Star OI™ involves people at all levels of an organization – not just management – and does so in an enjoyable way to hunt down areas of improvement and to prioritize solutions before implementation.

As the hunt plays out and the players move from starpoint to starpoint, they answer questions that help the organization stomp its crocs. Star OI™ is different, perhaps radically different, from other quality improvement methods because it:

- Ensures that all players throughout the organization are on the same page by using comprehensive communications
- Measures actual ongoing savings to the organization
- Establishes a structure to ensure that the savings are not only maintained, but increased, year after year

The unique value of Star OI™ is that it is independent of a particular method or standard, and can be integrated with any type of improvement initiative. If, for example, your company is approaching or at CMMI® Maturity Level 3 compliance, Star OI™ will make your progress to Maturity Level 4 and 5 easier and more effective.

Ralph Williams

No organization can eliminate all of its waste, loss, and inefficiency factors overnight. And you can be pretty sure that as some factors are eliminated, others will rise in their place. The difference is that with Star OI™ you will be much better able to deal with them, and they will, for the most part, be "smaller crocs." In truth, improvement must be an incremental effort, like the successive safaris in the story. The secondary challenge becomes keeping your crocs stomped as you eliminate the successive rounds of problems. We call this "holding your ground," just as Jason did in the Valley of Gold.

Star OI™ includes measures for documenting that you actually are holding your ground. These measures also translate into real dollar amounts saved. By holding your ground and adding up the savings over successive improvement efforts, your company will know the cumulative value of its savings as a result of using Star OI™.

The purpose behind the S*tomp Yer Croc*™ story is to show that improving your organization can be all-inclusive, fun, and rewarding for all involved. I believe that this book can point the way to revitalizing your business improvement efforts and bringing your organization to another level. I also believe there is a way to significantly improve your bottom line without selling more or reducing the work force.

I believe the gold is there for you, and I wish you every success and absolute joy on the hunt.

Ralph Williams
President
Cooliemon, LLC
January, 2008

**For more information,
and for assistance in finding your own Valley of Gold,
visit our website: www.cooliemon.com**

Stomp Yer Croc!

About the Author

RALPH WILLIAMS, president and founder of Cooliemon, LLC – a Carnegie Mellon University (CMU) Software Engineering Institute (SEI) Transition Partner – is one of the most active SEI CMMI® Lead Appraisers in the world. His unique ways of communication through storytelling, and his creative consulting style, have secured him a solid reputation in the business world and have made him an in-demand speaker at international conferences. Ralph earned his Bachelor's degree in Mathematics and Computer Science at Northumbria University (England).

Stomp Yer Croc!™ is the result of his extensive experience as a business and process improvement consultant for private and public sector organizations. His expertise in multiple models and standards, such as CMMI®, ITIL, ISO and IEEE – combined with practical working knowledge from classical quality improvement methods established by Juran, Deming, and Shewhart – have solidified his reputation in the organizational improvement field.

Ralph has provided services for clients ranging from such commercial firms as Raytheon, Northrop Grumman, Argon ST, Constella Health Sciences, and JP Morgan Chase Manhattan and government agencies such as the Federal Energy and Research Commission.

If you're ready to see your business grow, even beyond what you might have thought was possible, feel free to contact the Cooliemon Team. We'd be glad to show you how to find and **Stomp the Crocs** that stand between you and your goals!

And if you would like to buy additional copies of

Stomp Yer Croc!

for your management team – or your entire workforce – please contact us and ask about our generous quantity discounts.

Ralph Williams, President
Cooliemon, L.L.C.
www.cooliemon.com
e-mail: info@cooliemon.com